Thinking about

Gödel and Turing

Essays on Complexity, 1970 – 2007

Sans les mathématiques
on ne pénètre point au fond de la philosophie.
Sans la philosophie
on ne pénètre point au fond des mathématiques.
Sans les deux
on ne pénètre au fond de rien.

Leibniz

Thinking about
Gödel and Turing

Essays on Complexity, 1970 – 2007

Gregory J Chaitin

IBM T J Watson Research Center, USA

With a Foreword by **Paul Davies**

World Scientific

NEW JERSEY · LONDON · SINGAPORE · BEIJING · SHANGHAI · HONG KONG · TAIPEI · CHENNAI

Published by

World Scientific Publishing Co. Pte. Ltd.

5 Toh Tuck Link, Singapore 596224

USA office: 27 Warren Street, Suite 401-402, Hackensack, NJ 07601

UK office: 57 Shelton Street, Covent Garden, London WC2H 9HE

British Library Cataloguing-in-Publication Data
A catalogue record for this book is available from the British Library.

Author's photograph courtesy of Jacqueline Meyer.
For additional copyright information, see the Acknowledgements at the end of the book.

THINKING ABOUT GÖDEL AND TURING
Essays on Complexity, 1970–2007

ISBN-13 978-981-270-895-3
ISBN-10 981-270-895-2
ISBN-13 978-981-270-896-0 (pbk)
ISBN-10 981-270-896-0 (pbk)

Printed in Singapore.

Foreword

I was once asked to write an article about the most profound discovery ever made. I unhesitatingly chose Kurt Gödel's incompleteness theorem. In the normal use of the term, the word "discovery" implies that we learn something that was not known before. Gödel's theorem, by contrast, tells us what we *don't* know and can't know. It sets a fundamental and inescapable limit on knowledge of what is. It pinpoints the boundaries of ignorance — not just human ignorance, but that of any sentient being.

This remarkable book addresses the question of what can and cannot be known. It is about the nature of existence and reality and truth. Before Gödel, it was widely supposed that mathematics offered the most secure form of knowledge. Mathematics is a vast labyrinth of definitions and relationships open to independent scrutiny and supported by the iron scaffolding of unassailable logic. Human beings may legitimately quarrel about the facts of history or religion or politics, or even about the content of scientific theories, but properly-formulated mathematics leaves no scope for disagreement. The statement "eleven is a prime number" is not a matter of learned opinion, it is simply true, as may be demonstrated by systematic proof. That is to say, the statement is true because it can be *proved* to be true, step by step, starting with the assumed axioms of arithmetic and applying the standard rules of logic at each point in the argument. The end result is thus beyond any doubt. The utterly shocking import of Gödel's theorem, and the work of Emil Post and Alan Turing that flowed from it, is that the mighty edifice of mathematics is ultimately built on sand, because the nexus between proof and truth is demonstrably shaky. The problem that Gödel uncovered is that in mathematics, and in fact in almost all formal systems of reasoning, statements can be true yet unprovable — not just unproved, but *unprovable,* even in principle. Mathematical propositions can no longer be envisaged as a colossal list of statements to which yes-no answers may always be appended by exhaustive systematic investigation; rather, some of the propositions may be intrinsically undecidable, thus demolishing the concept of a closed, consistent and complete body of rules and objects. Incompleteness is unavoidable. The concept of *absolute truth,* even in the orderly world of mathematics, and even when that world is apprehended by a godlike intelligence, was dealt a shattering blow by Gödel's work.

When Gödel dropped his bombshell, the world of logic and rational argument were turned upside down, and it may have seemed to some, fleetingly, that science and mathematics would never be the same again. In the event, science, mathematics and daily life proceeded more or less normally in spite of it. To many scientists, Gödel's theorem was regarded as little more than an obscure curiosity, one that could be shrugged aside as largely irrelevant to the real world. Now, following the work of Gregory Chaitin, that is no longer a tenable response. Chaitin greatly extended the sweep of Gödel's basic insight, and re-cast the notion of incompleteness in a way that brings it much closer to the real world of computers and physical processes. A key step in his work is the recognition of a basic link between mathematical undecidability and randomness. Something is random if it has no pattern, no abbreviated description, in which case there is no algorithm shorter than the thing itself that captures its content. And a random fact is true for no reason at all; it is true "by accident," so to speak. With this conceptual tool-kit, Chaitin was able to demonstrate that mathematics is shot-through with randomness. It's there even in ordinary arithmetic! Mathematics, supposedly the epitome of logical orderliness, is exposed as harboring irreducible arbitrariness.

The implications for the physical world stem from the fact that the laws of physics are mathematical relationships: "The great book of Nature," proclaimed Galileo, "can be read only by those who know the language in which it was written. And this language is mathematics." The intimate relationship between mathematics and physics remains utterly mysterious, and touches on one of the deepest questions of physics and metaphysics, namely, where do the laws of physics come from? And why do they have the form that they do? In the orthodox view, which can be traced back at least to Newton, a fundamental dualism is posited to lie at the heart of nature. On the one hand there are immutable timeless universal and perfect mathematical laws, and on the other hand there are time-dependent contingent local states. A fundamental asymmetry connects them: the states evolve in a way that depends crucially on the laws of physics, but the laws of physics depend not one jot on the states.

Today, the prevailing view is that many of the laws of physics we find in the textbooks are actually only effective low-energy laws that emerged with the cooling of the universe from a hot big bang 13.7 billion years ago. Nevertheless, it is supposed that there must exist an underlying set of fundamental laws from which the low-energy laws derive, possibly with some random element brought about by symmetry breaking. A fashionable view is that string theory (or M theory) will provide these "ultimate" laws of physics. Whatever is the case, the conventional wisdom is that the fundamental laws are fixed and get imprinted on the universe from without — somehow! — at the moment of its birth (or, in more elaborate models such as eternal inflation, which have no ultimate origin, the laws are timelessly fixed). So physicists generally assume the laws of physics are transcendent. Furthermore, as I have stated, the laws are conventionally regarded as perfect, idealized mathematical relationships. Indeed, they explicitly incorporate idealized concepts such as real numbers and differentiability that require infinite and infinitesimal quantities. Thus theoretical physics today has a strongly Platonic flavor: the laws really exist as perfect, idealized, infinitely precise, timeless, immutable mathematical truths in some transcendent Platonic heaven. Furthermore, given that the laws of physics are a subset of all possible mathematical relationships, we have the image of Mother Nature plundering a vast warehouse of mathematics, plucking out a few choice items to employ as physical laws. So mathematics occupies a deeper, prior position in the Great Chain of Being. The physical universe is run by laws that could have been otherwise (that is, they are not required to have the form they have by dint of logical necessity), and so they belong to a contingent subset of mathematics. The mystery of how the choice is made — or what it is that "breaths fire into the equations and makes a universe for them to govern," to use Stephen Hawking's evocative phrase — remains unsolvable in this scheme.

Chaitin's work, however, calls into question the entire orthodox paradigm concerning the nature of physical law. "The basic insight," he writes, "is a software view of science: a scientific theory is like a computer program that predicts our observations, the experimental data." In other words, we may regard nature as an information processing system, and a law of physics as an algorithm that maps input data (initial conditions) into output data (final state). Thus in some sense the universe is a gigantic computer, with the laws playing the role of universal software. This shift in view is no mere semantic quibble. As Chaitin points out in what follows, mathematics contains randomness — or accidental, reasonless truths — because a computer, in the guise of a universal Turing machine, may or may not halt in executing its program, and there is no systematic way to know in advance if a function is computable (i.e. the Turing machine will halt) or not. But this raises an intriguing question. If the laws of physics are computer algorithms, will there also be randomness in physics stemming from Turing uncomputability? (I am not referring here to the well-known randomness inherent in quantum mechanics. The laws of quantum physics are not themselves random. Rather, they describe the precise evolution of states to which probability amplitudes are attached by hypothesis.) Well, I want to argue that not only is the answer yes, but that the randomness in the laws of physics is even more pronounced than that which flows from Turing uncomputability.

Let me give the gist of my argument. If the universe is a type of computer, then, like all man-made computers, it will be limited in its computational power. First, it has a finite information processing speed due to the fact that quantum mechanics imposes a lower limit to the transition time for flipping each bit of information. Second, the universe has a finite age, so a given finite volume of space will have processed only a finite amount of information since the big bang. Thirdly, the finite speed of light implies that the volume of space to which we have causal access at this time is finite, so there is a finite amount of information storage available. (The same conclusion may be arrived at, incidentally, by invoking the so-called holographic principle, a derivative of the well-known Bekenstein-Hawking result that the information content of a black hole is proportional to its surface area.) The real universe therefore differs in a crucial respect from the concept of a Turing machine. The latter is supposed to have infinite time at its disposal: there is no upper bound on the number of steps it may perform to execute its program. The only relevant issue is whether the program eventually halts or not, however long it takes. The machine is also permitted unlimited memory, in the form of an unbounded paper tape. If these limitless resources are replaced by *finite* resources, however, an additional, fundamental, source of unknowability emerges. So if, following Chaitin, we treat the laws of physics as software running on the resource-limited hardware known as the observable universe, then these laws will embed a form of randomness, or uncertainty, or ambiguity, or fuzziness — call it what you will — arising from the finite informational processing capacity of the cosmos. Thus there will be a *cosmological* bound on the fidelity of all mathematical laws.

To accept my argument you have to make a choice concerning the nature of mathematical laws. A dedicated Platonist can dismiss the finite resource issue by claiming that Mother Nature cares nothing for the computational limitations of the real universe because she computes happily in the infinitely-resourced, infinitely precise, timeless Platonic heaven, and merely delivers the output to the physical universe in its exact and completed form. But an alternative view, first entertained thirty years ago by John Wheeler and Rolf Landauer, is that the laws of physics are not transcendent *of* the physical universe, but inherent *in* it, and emergent with it at the big bang. Landauer and Wheeler were motivated by their belief that information, as opposed to matter, is the fundamental currency of nature, a viewpoint summed up in Wheeler's well-known aphorism "It from bit!". Landauer went on to postulate a new physical principle, namely: *A physical theory should not invoke calculative routines that in fact cannot be carried out.* The finiteness of the real universe has a clear implication for Landauer's principle, as he himself explicated: "The calculative process, just like the measurement process, is subject to some limitations. A sensible theory of physics must respect these limitations."

For Landauer and Wheeler, mathematics and physics stand in a symmetric relationship, forming a self-consistent loop: the laws of the universe are mathematical computations, but mathematics (at least as manifested in physical law) is what the universe computes. Thus mathematical law and physics co-emerge. The link between mathematics and physics, which in the orthodox view is asymmetric and unexplained, is given the basis of an explanation. Let me repeat that the randomness or fuzziness of the laws of physics to which I refer is not quantum uncertainty, but an irreducible unknowability in the laws themselves. The scale of the imprecision is given not by Planck's constant, but by a cosmological parameter related to the size of the universe. For this reason, the bound was lower, and hence more stringent, in the past.

Today, the information bound is approximately 10^{122} bits, but at the time of cosmological inflation it may have been as small as 10^{20}. We can imagine the laws as emerging in a fuzzy and ill-defined manner at the birth of the cosmos, and then focusing down on the observed set as the universe expands, ages, and envelops greater and greater computational resources. During the very early universe, the information bound may have been severe enough to affect the formation of the large-scale structure of the universe, and to leave an observational signature in the cosmic microwave background.

Chaitin's work has exposed the essential unknowability of idealized mathematical objects, for example, (almost all) real numbers. But in a finitely-resourced universe, the situation is far worse. Consider, for example, a generic entangled state of 500 quantum particles. To specify this state, one requires a list of $2^{500} \sim 3 \times 10^{150}$ amplitudes, one for each branch of the wavefunction. It is physically impossible, even in principle, to give these numbers, because they far exceed the theoretical maximum information capacity of the entire observable universe. So the quantum state cannot be specified, let alone its unitary evolution predicted. What justification is there, then, for the claim that the laws of quantum mechanics will unfailingly describe this unknowable state and its equally unknowable evolution? It is worth noting that although practical quantum entanglement is so far limited to a dozen or so particles, the target for a forthcoming quantum computer is in the region of 10,000 entangled particles. This does not necessarily mean that a quantum computer will not work for calculations of interest (e.g. prime factoring), because many problems in mathematics may be solvable using highly algorithmically compressed specifications of the initial state, and the state vector might evolve within only a tiny subspace of the entire $2^{10,000}$-dimensional Hilbert space, although to my knowledge there is no general proof of this.

How do these deliberations relate to the problem of the meaning of existence and computability? It is possible to maintain that "large" computable functions still exist, in some abstract sense, even though they may not be computable in the real physical universe. In other words, mathematical existence can be tied to idealized Turing computability as opposed to cosmic computability. But what is clear is that if one accepts Landauer's principle that only cosmic-computable functions should be invoked to describe the real physical cosmos, then the distinction between Turing and cosmic computability could lead to definite and potentially observable consequences. So my conclusions are very much in accordance with Chaitin's bold claim that "perhaps mathematics and physics are not as different as most people think."

All historical eras have their metaphors for the universe. In ancient Greece, musical instruments and surveying equipment represented the pinnacle of technology, and the Greek philosophers built a cosmology based on harmony and geometrical perfection. In Newton's day, the clockwork was the technological marvel of the age, and Newton gave us the clockwork universe. Two centuries later the thermodynamic universe followed from the development of the steam engine. Today, the digital computer is the technology we find most dazzling, and we now recognize that the cosmos is a vast information processing system. Each new era brings fresh insights into the nature of the physical world. This book describes the most ambitious attempt yet in the history of human thought to grapple with the deep nature of reality. Among the many big questions of existence so far posed by mankind, they don't come much bigger than this.

PAUL DAVIES
Beyond: Center for Fundamental Concepts in Science
Arizona State University, Tempe, Arizona

Preface

This year will be my 60th birthday, which naturally makes one look back to see what one has accomplished and failed to accomplish. This is also an age at which the technical details begin to seem less important and the big, philosophical ideas stand out.

For four decades I have been using the idea of complexity to try to understand the significance of Gödel's famous incompleteness theorem. Most logicians and mathematicians think that this theorem can be dismissed. I, on the contrary, believe that mathematics must be conceived of and carried out differently because of Gödel.

The fundamental question: What is a proof? Why is it convincing? What is mathematics? In fact, Gödel's proof is a *reductio ad absurdum* of the idea of a formal axiomatic math theory. Gödel, Turing and myself, what we do each in our own unique way, is to assert increasingly emphatically that math is not a formal theory, it is **not** mechanical. What then **is** math? Where do new truths come from?

Instead of saying what math isn't, how about a new, optimistic metamathematics that says what math is, and how creativity, imagination and inspiration make it progress? For like any living organism, math must progress or it withers and dies. It cannot be static, it must be dynamic, it must constantly evolve. And there are other pressing questions, particularly for those of us who feel so attracted, so obsessed by mathematics. We sometimes wonder: Why is it so beautiful? What precisely attracts us? Why do we feel so attracted? Is there no end to such beauty?

This collection contains twenty-three papers that I still feel to be stimulating and which discuss these issues at a philosophical rather than at a technical level. They are presented in chronological order, in order of publication, and taken together I hope they begin to make a case for a new way of looking at mathematics, for a new idea of what math is all about, more in line with how physicists work, perhaps.

This is, I hope, a case in which the whole is greater than the sum of its parts. Many interesting topics are discussed, including Cantor's diagonal method, Gödel's 1931 proof, Turing's halting problem, program-size complexity, algorithmic irreducibility and randomness, as well as important ideas on complexity and the limits of mathematics of Leibniz and Emile Borel that I feel are insufficiently appreciated.

By going through this collection in chronological order one can also appreciate how difficult it is for new ideas to emerge, as one slowly gropes in the dark for understanding. One keeps turning things over and over in one's mind. Each tiny step forward takes years. One only perseveres because one must and because one believes that things ultimately make sense, even if one will never fully understand why.

I also hope that this collection, showing as it does how a few simple ideas, mere seedlings at first, gradually developed into a large tree, will encourage others to believe in the power of ideas, in math as an art, and in the search for deep understanding and fundamental ideas.

Just look at those magic creative moments in human history, like ancient Greece, the Italian renaissance, France before the revolution, Vienna between the wars, when there is enough anarchy for new ideas to flourish, and individuals can affect events. Can we understand the sociodynamics of this magic? Can we encourage and promote such moments?

The basic lesson that I have learned from Gödel is that mathematics is not a machine. Creativity is essential. And it is also mysterious. Just look at the ease with which rivers of beautiful mathematics flowed from Euler's pen. Look at Ramanujan's remark that a goddess brought him ideas while he slept, and that no equation is worthwhile unless it expresses one of God's thoughts.[1] Or, for that matter, look at Gödel's faith that mathematicians can overcome the incompleteness theorem by intuiting new concepts and principles whenever this is needed for mathematics to advance. So it is high time for us to give up on static, formal mathematics and instead begin to study creativity and how ideas evolve.

Finally, I want to say that I am very grateful to my friends in Chile, Eric Goles, Oscar Orellana, and Ricardo Espinoza, for encouraging me to put together this collection and helping me to find an appropriate title.

GREGORY CHAITIN, Viña del Mar, January 2007

---

[1] In this connection, see Ira Hauptman's play *Partition* about Hardy, Ramanujan, and Ramanujan's goddess Namagiri. Available from `http://www.playscripts.com`.

Contents

Introductory note

How should this book be read? Well, the articles in it are independent, self-contained pieces, and I prefer to let readers wander through, having their own thoughts, exploring on their own, rather than offer a guided tour. In other words, I will let the individual essays stand on their own, unintroduced. And there is no need to read this book from cover to cover. Just read whatever strikes your fancy, enjoy whatever catches your eye.

However, if you do read this book from cover to cover in chronological order, you will see that the papers in it all deal with the same problem, they attempt to answer the same question: "What is the meaning of Gödel's incompleteness theorem?" Of course, my point of view changes and develops over time. Themes enter and disappear, but there is a central spine that never varies, a single thread that ties it all together. It's one train of thought, on different aspects of the same topic.

For those of you who would like a historical perspective, I have in fact put together a timeline explaining the evolution of my ideas. It's called "Algorithmic information theory: Some recollections." This, however, is a technical paper, not a popular account intended for the general reader. This timeline can be found in the *festschrift* volume assembled by Cristian Calude, *Randomness and Complexity, from Leibniz to Chaitin* (World Scientific, 2007).

The original sources of the papers in this collection are given in the table of contents, but more detailed information, including copyrights, appears in the **Acknowledgements** at the end of the book. And for those of you who would like to know where to go for more information on particular topics, I have included a **List of publications** with most of my technical and non-technical papers and books and some interviews.

1

On the difficulty of computations

Two practical considerations concerning the use of computing machinery are the amount of information that must be given to the machine for it to perform a given task and the time it takes the machine to perform it. The size of programs and their running time are studied for mathematical models of computing machines. The study of the amount of information (i.e., number of bits) in a computer program needed for it to put out a given finite binary sequence leads to a definition of a random sequence; the random sequences of a given length are those that require the longest programs. The study of the running time of programs for computing infinite sets of natural numbers leads to an arithmetic of computers, which is a distributive lattice. [This paper was presented at the Pan-American Symposium of Applied Mathematics, Buenos Aires, Argentina, August 1968.]

Section I

The modern computing machine sprang into existence at the end of World War II. But already in 1936 Turing and Post had proposed a mathematical model of computing machines (figure 1).[1] The mathematical model of the computing machine that Turing and Post proposed, commonly referred to as the Turing machine, is a black box with a finite number of internal states. The box can read and write on an infinite paper tape, which is divided into squares. A digit or letter may be written on each square of the tape, or the square may be blank. Each second the machine performs one of the following

---

[1]Their papers appear in Davis [1]. As general references on computability theory we may also cite Davis [2]–[4], Minsky [5], Rogers [6], and Arbib [7].

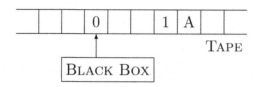

Figure 1. A Turing-Post machine

actions. It may stop, it may shift the tape one square to the right or one square to the left, it may erase the square on which the read-write head is positioned, or it may write a digit or letter on the square on which the read-write head is positioned. The action it performs is determined solely by the internal state of the black box at the moment, and the current state of the black box is determined solely by its previous internal state and the character read on the square of the tape on which its read-write head was positioned.

Incredible as it may seem at first, a machine of such primitive design can multiply numbers written on its tape, and can write on its tape the successive digits of π. Indeed, it is now generally accepted that any calculation that a modern electronic digital computer or a human computer can do, can also be done by such a machine.

Section II

How much information must be provided to a computer in order for it to perform a given task? The point of view we will present here is somewhat different from the usual one. In a typical scientific application, the computer may be used to analyze statistically huge amounts of data and produce a brief report in which a great many observations are reduced to a handful of statistical parameters. We would view this in the following manner. The same final result could have been achieved if we had provided the computer with a table of the results, together with instructions for printing them in a neat report. This observation is, of course, ridiculous for all practical purposes. For, had we known the results, it would not have been necessary to use a computer. This example, then, does not exemplify those aspects of computation that we will emphasize.

Rather, we are thinking of such scientific applications as solving the Schrödinger wave equation for the helium atom. Here we have no data, only a program; and the program will produce after much calculation a great deal of printout. Or consider calculating the apparent positions of the planets as observed from the earth over a period of years. A small program incorporating the very simple Newtonian theory for this situation will predict a great many astronomical observations. In this problem there are no data—only a program that contains, of course, a table of the masses of the planets and their initial positions and velocities.

Section III

Let us now consider the problem of the amount of information that it is necessary to provide to a computer in order for it to calculate a given finite binary sequence. A computing machine is defined for these purposes to be a device that accepts as input a program, performs the calculations indicated to it in the program, and finally puts out the binary sequence it has calculated. In line with the mathematical theory of information, it is natural for the program to be viewed as a sequence of bits or 0's and 1's. Furthermore, in computer engineering all programs and data are represented in the machine's circuits in binary form. Thus, we may consider a computer to be a device that accepts one binary sequence (the program) and emits another (the result of the calculation).

$$011001001 \rightarrow \text{COMPUTER} \rightarrow 1111110010001100110100$$

As an example of a computer we would then have an electronic digital computer that accepts programs consisting of magnetized spots on magnetic tape and puts out its results in the same form. Another example is a Turing machine. The program is a series of 0's and 1's written on the machine's tape at the start of the calculation, and the result is a sequence of 0's and 1's written on its tape when it stops. As was mentioned, the second of these examples can do anything that the first can.

Section IV

We are interested in the amount of information that must be supplied to a computer M in order for it to calculate a given finite binary sequence S. We may now define this as the size or length of the smallest binary sequence that causes the machine M to calculate S. We denote the length of the shortest program for M to calculate S by $L(M, S)$. It has been shown that there is a computing machine M that has the following three properties.[2]

    1) $L(M, S) \leq k + 1$ for all binary sequences S of length k.

In other words, any binary sequence of length k can be calculated by this computer M if it is given an appropriate program at most $k + 1$ bits in length. The proof is as follows. If no better way to calculate a binary sequence occurs to us, we can always include the binary sequence as a table in the program. This computer is so designed that we need add only a single bit to the sequence to obtain a program for computing it. The computer M emits the sequence S when it is given the program $S0$.

    2) Those binary sequences S for which $L(M, S) < j$ are fewer than 2^j in number.

Thus, most binary sequences of length k require programs of about the same length k, and the number of sequences that can be computed by smaller programs decreases exponentially as the size of the program decreases. The proof is as follows. There are only $2^j - 2$ binary sequences less than j in length. Thus, there are fewer than 2^j programs less than j in length, for each program is a binary sequence. At best, a program will cause the computer to calculate a single binary sequence. At worst, an error in the program will trap the computer in an endless loop, and no binary sequence will be calculated. As each program causes the computer to calculate at most one binary sequence, the number of sequences calculated must be smaller than the number of programs. Thus, fewer than 2^j binary sequences can be calculated by means of programs less than j in length.

    3) For any other computer M' there exists a constant $c(M')$ such that for all binary sequences S, $L(M, S) \leq L(M', S) + c(M')$.

---

[2]Solomonoff [8] was the first to employ computers of this kind.

In other words, this computer requires shorter programs than any other computer, or more exactly it does not require programs much longer than those required by any other computer. The proof is as follows. The computer M is designed to interpret the circuit diagrams of any other computer M'. Given a program for M' and the circuit diagrams of M', the computer M proceeds to calculate how M' would behave, i.e., it proceeds to simulate M'. Thus, we need only add a fixed number of bits to any program for M' in order to obtain a program that enables M to calculate the same result. This program for M is of the form $PC1$.

The 1 at the right end of the program indicates to the computer M that this is a simulation, C is a fixed binary sequence of length $c(M') - 1$ giving the circuit diagrams of the computer M', which is to be imitated, and P is the program for M'.[3]

Section V

Kolmogorov [9] and the author [11], [12] have independently suggested that computers such as those previously described be applied to the problem of defining what is meant by a random or patternless finite binary sequence of 0's and 1's. In the traditional foundations of the mathematical theory of probability, as expounded by Kolmogorov in his classic [10], there is no place for the concept of an individual random sequence of 0's and 1's. Yet it is not altogether meaningless to say that the sequence

$$11001011110011001011110000010$$

is more random or patternless than the sequences

$$111111111111111111111111111111$$
$$010101010101010101010101010101,$$

for we may describe these last two sequences as thirty 1's or fifteen 01's, but there is no shorter way to specify the first sequence than by just writing it all out.

We believe that the random or patternless sequences of a given length are those that require the longest programs. We have seen that most of the

---

[3]How can the computer M separate PC into P and C? C has each of its bits doubled, except the pair of bits at its left end. These are unequal and serve as punctuation separating C from P.

binary sequences of length k require programs of about length k. These, then, are the random or patternless sequences. Those sequences that can be obtained by putting into a computer a program much shorter than k are the nonrandom sequences, those that possess a pattern or follow a law. The more possible it is to compress a binary sequence into a short program calculation, the less random is the sequence.

As an example of this, let us consider those sequences of 0's and 1's in which 0's and 1's do not occur with equal frequency. Let p be the relative frequency of 1's, and let $q = 1 - p$ be the relative frequency of 0's. A long binary sequence that has the property that 1's are more frequent than 0's can be obtained from a computer program whose length is only that of the desired sequence reduced by a factor $H(p, q) = -p \log_2 p - q \log_2 q$. For example, if 1's occur approximately $\frac{3}{4}$ of the time and 0's occur $\frac{1}{4}$ of the time in a long binary sequence of length k, there is a program for computing that sequence with length only about $H(\frac{3}{4}, \frac{1}{4})k = 0.80k$. That is, the program need be only approximately 80 percent the length of the sequence it computes. In summary, if 0's and 1's occur with unequal frequencies, we can compress such sequences into programs only a certain percentage (depending on the frequencies) of the size of the sequence. Thus, random or incompressible sequences will have about as many 0's as 1's, which agrees with our intuitive expectations.

In a similar manner it can be shown that all groups of 0's and 1's will occur with approximately the expected frequency in a long binary sequence that we call random; 01100 will appear $2^{-5}k$ times in long sequences of length k, etc.[4]

Section VI

The definition of random or patternless finite binary sequences just presented is related to certain considerations in information theory and in the methodology of science.

The two problems considered in Shannon's classical exposition [15] are to transmit information as efficiently and as reliably as possible. Here we are interested in examining the viewpoint of information theory concerning the efficient transmission of information. An information source may be redundant, and information theory teaches us to code or compress messages

---

[4]Martin-Löf [14] also discusses the statistical properties of random sequences.

so that what is redundant is eliminated and communications equipment is optimally employed. For example, let us consider an information source that emits one symbol (either an A or a B) each second. Successive symbols are independent, and A's are three times more frequent than B's. Suppose it is desired to transmit the messages over a channel that is capable of transmitting either an A or a B each second. Then the channel has a capacity of 1 bit per second, while the information source has entropy 0.80 bits per symbol; and thus it is possible to code the messages in such a way that on the average $1/0.80 = 1.25$ symbols of message are transmitted over the channel each second. The receiver must decode the messages; that is, expand them into their original form.

In summary, information theory teaches us that messages from an information source that is not completely random (that is, which does not have maximum entropy) can be compressed. The definition of randomness is merely the converse of this fundamental theorem of information theory; if lack of randomness in a message allows it to be coded into a shorter sequence, then the random messages must be those that cannot be coded into shorter messages. A computing machine is clearly the most general possible decoder for compressed messages. We thus consider that this definition of randomness is in perfect agreement and indeed strongly suggested by the coding theorem for a noiseless channel of information theory.

Section VII

This definition is also closely related to classical problems of the methodology of science.[5]

Consider a scientist who has been observing a closed system that once every second either emits a ray of light or does not. He summarizes his observations in a sequence of 0's and 1's in which a 0 represents "ray not emitted" and a 1 represents "ray emitted." The sequence may start

$$0110101110\ldots$$

and continue for a few million more bits. The scientist then examines the sequence in the hope of observing some kind of pattern or law. What does he mean by this? It seems plausible that a sequence of 0's and 1's is patternless

---

[5]Solomonoff [8] also discusses the relation between program lengths and the problem of induction.

if there is no better way to calculate it than just by writing it all out at once from a table giving the whole sequence. The scientist might state:

My Scientific Theory: 0110101110...

This would not be considered an acceptable theory. On the other hand, if the scientist should hit upon a method by which the whole sequence could be calculated by a computer whose program is short compared with the sequence, he would certainly not consider the sequence to be entirely patternless or random. The shorter the program, the greater the pattern he may ascribe the sequence.

There are many parallels between the foregoing and the way scientists actually think. For example, a simple theory that accounts for a set of facts is generally considered better or more likely to be true than one that needs a large number of assumptions. By "simplicity" is not meant "ease of use in making predictions." For although general relativity is considered to be the simple theory par excellence, very extended calculations are necessary to make predictions from it. Instead, one refers to the number of arbitrary choices that have been made in specifying the theoretical structure. One is naturally suspicious of a theory whose number of arbitrary elements is of an order of magnitude comparable to the amount of information about reality that it accounts for.

Section VIII

Let us now turn to the problem of the amount of time necessary for computations.[6] We will develop the following thesis. Call an infinite set of natural numbers perfect if there is no essentially quicker way to compute infinitely many of its members than computing the whole set. Perfect sets exist. This thesis was suggested by the following vague and imprecise considerations.[7]

One of the most profound problems of the theory of numbers is that of calculating large primes. While the sieve of Eratosthenes appears to be as quick an algorithm for calculating all the primes as is possible, in recent times hope has centered on calculating large primes by calculating a subset

---

[6]As general references we may cite Blum [16] and Arbib and Blum [17]. Our exposition is a summary of that of [13].

[7]See Hardy and Wright [18], Sections 1.4 and 2.5 for the number-theoretic background of the following remarks.

of the primes, those that are Mersenne numbers. Lucas's test can decide the primality of a Mersenne number with rapidity far greater than is furnished by the sieve method. If there are an infinity of Mersenne primes, then it appears that Lucas has achieved a decisive advance in this classical problem of the theory of numbers.

An opposing point of view is that there is no essentially better way to calculate large primes than by calculating them all. If this is the case, it apparently follows that there must be only finitely many Mersenne primes.

These considerations, then, suggested that there are infinite sets of natural numbers that are arbitrarily difficult to compute, and that do not have any infinite subsets essentially easier to compute than the whole set. Here difficulty of computation refers to speed. Our development will be as follows. First, we define computers for calculating infinite sets of natural numbers. Then we introduce a way of comparing the rapidity of computers, a transitive binary relation, i.e., almost a partial ordering. Next we focus our attention on those computers that are greater than or equal to all others under this ordering, i.e., the fastest computers. Our results are conditioned on the computers having this property. The meaning of "arbitrarily difficult to compute" is then clarified. Last, we exhibit sets that are arbitrarily difficult to compute and do not have any subset essentially easier to compute than the whole set.

Section IX

We are interested in the speed of programs for generating the elements of an infinite set of natural numbers. For these purposes we may consider a computer to be a device that once a second emits a (possibly empty) finite set of natural numbers and that once started never stops. That is to say, a computer is now viewed as a function whose arguments are the program and the time and whose value is a finite set of natural numbers. If a program causes the computer to emit infinitely many natural numbers in size order and without any repetitions, we say that the computing machine calculates the infinite set of natural numbers that it emits.

A Turing machine can be used to compute infinite sets of natural numbers; it is only necessary to establish a convention as to when natural numbers are emitted. For example, we may divide the machine's tape into two halves, and stipulate that what is written on the right half cannot be erased.

The computational scratchwork is done on the left half of the tape, and the successive members of the infinite set of natural numbers are written on the nonerasable squares in decimal notation, separated by commas, with no blank spaces permitted between characters. The moment a comma has been written, it is considered that the digits between it and the previous comma form the numeral representing the next natural number emitted by the machine. We suppose that the Turing machine performs a single cycle of activity (read tape; shift, write, or erase tape; change internal state) each second. Last, we stipulate that the machine be started scanning the first nonerasable square of the tape, that initially the nonerasable squares be all blank, and that the program for the computer be written on the first erasable squares, with a blank serving as punctuation to indicate the end of the program and the beginning of an infinite blank region of tape.

Section X

We now order the computers according to their speeds. $C \geq C'$ is defined as meaning that C is not much slower than C'.

What do we mean by saying that computer C is not much slower than computer C' for the purpose of computing infinite sets of natural numbers? There is a computable change of C's time scale that makes C as fast as C' or faster. More exactly, there is a computable function $f(n)$ (for example $n!$ or $n^{n^{n^{\cdots}}}$ with n exponents) with the following property. Let P' be any program that makes C' calculate an infinite set of natural numbers. Then there exists a program P that makes C calculate the same set of natural numbers and has the additional property that every natural number emitted by C' during the first t seconds of calculation is emitted by C during the first $f(t)$ second of calculation, for all but a finite number of values of t. We may symbolize this relation between the computers C and C' as $C \geq C'$, for it has the property that $C \geq C'$ and $C' \geq C''$ only if $C \geq C''$.

In this way, we have introduced an ordering of the computers for computing infinite sets of natural numbers, and it can be shown that a distributive lattice results. The most important property of this ordering for our present purposes is that there is a set of computers $\geq$ all other computers. In what follows we assume that the computer that is used is a member of this set of fastest computers.

Section XI

We now clarify what we mean by "arbitrarily difficult to compute."

Let $f(n)$ be any computable function that carries natural numbers into natural numbers. Such functions can get big very quickly indeed. For example consider the function $n^{n^{n^{\cdots}}}$ in which there are n^n exponents. There are infinite sets of natural numbers such that, no matter how the computer is programmed, at least $f(n)$ seconds will pass before the computer emits all those elements of the set that are less than or equal to n. Of course, a finite number of exceptions are possible, for any finite part of an infinite set can be computed very quickly by including in the computer's program a table of the first few elements of the set. Note that the difficulty in computing such sets of natural numbers does not lie in the fact that their elements get very big very quickly, for even small elements of such sets require more than astronomical amounts of time to be computed. What is more, there are infinite sets of natural numbers that are arbitrarily difficult to compute and include 90 percent of the natural numbers.

We finally exhibit infinite sets of natural numbers that are arbitrarily difficult to compute, and do not have any infinite subsets essentially easier to compute than the whole set. Consider the following tree of natural numbers (figure 2).[8] The infinite sets of natural numbers that we promised to exhibit are obtained by starting at the root of the tree (that is, at 0) and walking forward, including in the set every natural number that is stepped on.

It is easy to see that no infinite subset of such a set can be computed much more quickly than the whole set. For suppose we are told that n is in such a set. Then we know at once that the greatest integer less than $n/2$ is the previous element of the set. Thus, knowing that 1 000 000 is in the set, we immediately produce all smaller elements in it, by walking backwards through the tree. They are 499 999, 249 999, 124 999, etc. It follows that there is no appreciable difference between generating an infinite subset of such a set, and generating the whole set, for gaps in an incomplete generation can be filled in very quickly.

It is also easy to see that there are sets that can be obtained by walking through this tree and are arbitrarily difficult to compute. These, then, are the sets that we wished to exhibit.

---

[8]This tree is used in Rogers [6], p. 158, in connection with retraceable sets. Retraceable sets are in some ways analogous to those sets that concern us here.

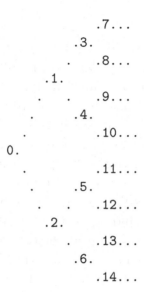

Figure 2. A tree of natural numbers

Acknowledgment

The author wishes to express his gratitude to Prof. G. Pollitzer of the University of Buenos Aires, whose constructive criticism much improved the clarity of this presentation.

References

[1] M. Davis, Ed., *The Undecidable.* Hewlett, N.Y.: Raven Press, 1965.

[2] —, *Computability and Unsolvability.* New York: McGraw-Hill, 1958.

[3] —, "Unsolvable problems: A review," *Proc. Symp. on Mathematical Theory of Automata.* Brooklyn, N.Y.: Polytech. Inst. Brooklyn Press, 1963, pp. 15–22.

[4]   , "Applications of recursive function theory to number theory," *Proc. Symp. in Pure Mathematics,* vol. 5. Providence, R.I.: AMS, 1962, pp. 135–138.

[5] M. Minsky, *Computation: Finite and Infinite Machines.* Englewood Cliffs, N.J.: Prentice-Hall, 1967.

[6] H. Rogers, Jr., *Theory of Recursive Functions and Effective Computability.* New York: McGraw-Hill, 1967.

[7] M. A. Arbib, *Theories of Abstract Automata*. Englewood Cliffs, N.J.: Prentice-Hall (to be published).

[8] R. J. Solomonoff, "A formal theory of inductive inference," *Inform. and Control*, vol. 7, pp. 1–22, March 1964; pp. 224–254, June 1964.

[9] A. N. Kolmogorov, "Three approaches to the definition of the concept 'quantity of information'," *Probl. Peredachi Inform.*, vol. 1, pp. 3–11, 1965.

[10] —, *Foundations of the Theory of Probability*. New York: Chelsea, 1950.

[11] G. J. Chaitin, "On the length of programs for computing finite binary sequences," *J. ACM*, vol. 13, pp. 547–569, October 1966.

[12] —, "On the length of programs for computing finite binary sequences: statistical considerations," *J. ACM*, vol. 16, pp. 145–159, January 1969.

[13] —, "On the simplicity and speed of programs for computing infinite sets of natural numbers," *J. ACM*, vol. 16, pp. 407–422, July 1969.

[14] P. Martin-Löf, "The definition of random sequences," *Inform. and Control*, vol. 9, pp. 602–619, December 1966.

[15] C. E. Shannon and W. Weaver, *The Mathematical Theory of Communication*. Urbana, Ill.: University of Illinois Press, 1949.

[16] M. Blum, "A machine-independent theory of the complexity of recursive functions," *J. ACM*, vol. 14, pp. 322–336, April 1967.

[17] M. A. Arbib and M. Blum, "Machine dependence of degrees of difficulty," *Proc. AMS*, vol. 16, pp. 442–447, June 1965.

[18] G. H. Hardy and E. M. Wright, *An Introduction to the Theory of Numbers*. Oxford: Oxford University Press, 1962.

The following references have come to the author's attention since this lecture was given.

[19] D. G. Willis, "Computational complexity and probability constructions," Stanford University, Stanford, Calif., March 1969.

[20] A. N. Kolmogorov, "Logical basis for information theory and probability theory," *IEEE Trans. Information Theory*, vol. IT-14, pp. 662–664, September 1968.

[21] D. W. Loveland, "A variant of the Kolmogorov concept of complexity," Dept. of Math., Carnegie-Mellon University, Pittsburgh, Pa., Rept. 69-4.

[22] P. R. Young, "Toward a theory of enumerations," *J. ACM*, vol. 16, pp. 328–348, April 1969.

[23] D. E. Knuth, *The Art of Computer Programming; vol. 2, Seminumerical Algorithms*. Reading, Mass.: Addison-Wesley, 1969.

[24] *1969 Conf. Rec. of the ACM Symp. on Theory of Computing* (Marina del Rey, Calif.).

Information-theoretic computational complexity

This paper attempts to describe, in nontechnical language, some of the concepts and methods of one school of thought regarding computational complexity. It applies the viewpoint of information theory to computers. This will first lead us to a definition of the degree of randomness of individual binary strings, and then to an information-theoretic version of Gödel's theorem on the limitations of the axiomatic method. Finally, we will examine in the light of these ideas the scientific method and von Neumann's views on the basic conceptual problems of biology. [This paper was presented at the IEEE International Congress of Information Theory, Ashkelon, Israel, June 1973.]

This field's fundamental concept is the complexity of a binary string, that is, a string of bits, of zeros and ones. The complexity of a binary string is the minimum quantity of information needed to define the string. For example, the string of length n consisting entirely of ones is of complexity approximately $\log_2 n$, because only $\log_2 n$ bits of information are required to specify n in binary notation.

However, this is rather vague. Exactly what is meant by the definition of a string? To make this idea precise a computer is used. One says that a string defines another when the first string gives instructions for constructing the second string. In other words, one string defines another when it is a program for a computer to calculate the second string. The fact that a string of n ones is of complexity approximately $\log_2 n$ can now be translated more correctly into the following. There is a program $\log_2 n + c$ bits long that calculates the string of n ones. The program performs a loop for printing ones n times. A fixed number c of bits are needed to program the loop, and $\log_2 n$ bits more for specifying n in binary notation.

Exactly how are the computer and the concept of information combined to define the complexity of a binary string? A computer is considered to take one binary string and perhaps eventually produce another. The first string is the program that has been given to the machine. The second string is the output of this program; it is what this program calculates. Now consider a given string that is to be calculated. How much information must be given to the machine to do this? That is to say, what is the length in bits of the shortest program for calculating the string? This is its complexity.

It can be objected that this is not a precise definition of the complexity of a string, inasmuch as it depends on the computer that one is using. Moreover, a definition should not be based on a machine, but rather on a model that does not have the physical limitations of real computers.

Here we will not define the computer used in the definition of complexity. However, this can indeed be done with all the precision of which mathematics is capable. Since 1936 it has been known how to define an idealized computer with unlimited memory. This was done in a very intuitive way by Turing and also by Post, and there are elegant definitions based on other principles [2]. The theory of recursive functions (or computability theory) has grown up around the questions of what is computable and what is not.

Thus it is not difficult to define a computer mathematically. What remains to be analyzed is which definition should be adopted, inasmuch as some computers are easier to program than others. A decade ago Solomonoff solved this problem [7]. He constructed a definition of a computer whose programs are not much longer than those of any other computer. More exactly, Solomonoff's machine simulates running a program on another computer, when it is given a description of that computer together with its program.

Thus it is clear that the complexity of a string is a mathematical concept, even though here we have not given a precise definition. Furthermore, it is a very natural concept, easy to understand for those who have worked with computers. Recapitulating, the complexity of a binary string is the information needed to define it, that is to say, the number of bits of information that must be given to a computer in order to calculate it, or in other words, the size in bits of the shortest program for calculating it. It is understood that a certain mathematical definition of an idealized computer is being used, but it is not given here, because as a first approximation it is sufficient to think of the length in bits of a program for a typical computer in use today.

Now we would like to consider the most important properties of the complexity of a string. First of all, the complexity of a string of length n is less than $n + c$, because any string of length n can be calculated by putting it directly into a program as a table. This requires n bits, to which must be added c bits of instructions for printing the table. In other words, if nothing betters occurs to us, the string itself can be used as its definition, and this requires only a few more bits than its length.

Thus the complexity of each string of length n is less than $n+c$. Moreover, the complexity of the great majority of strings of length n is approximately n, and very few strings of length n are of complexity much less than n. The reason is simply that there are much fewer programs of length appreciably less than n than strings of length n. More exactly, there are 2^n strings of length n, and less than 2^{n-k} programs of length less than $n - k$. Thus the number of strings of length n and complexity less than $n - k$ decreases exponentially as k increases.

These considerations have revealed the basic fact that the great majority of strings of length n are of complexity very close to n. Therefore, if one generates a binary string of length n by tossing a fair coin n times and noting whether each toss gives head or tail, it is highly probable that the complexity of this string will be very close to n. In 1965 Kolmogorov proposed calling random those strings of length n whose complexity is approximately n [8]. We made the same proposal independently [9]. It can be shown that a string that is random in this sense has the statistical properties that one would expect. For example, zeros and ones appear in such strings with relative frequencies that tend to one-half as the length of the strings increases.

Consequently, the great majority of strings of length n are random, that is, need programs of approximately length n, that is to say, are of complexity approximately n. What happens if one wishes to show that a particular string is random? What if one wishes to prove that the complexity of a certain string is almost equal to its length? What if one wishes to exhibit a specific example of a string of length n and complexity close to n, and assure oneself by means of a proof that there is no shorter program for calculating this string?

It should be pointed out that this question can occur quite naturally to a programmer with a competitive spirit and a mathematical way of thinking. At the beginning of the sixties we attended a course at Columbia University in New York. Each time the professor gave an exercise to be programmed, the students tried to see who could write the shortest program. Even though several times it seemed very difficult to improve upon the best program that had been discovered, we did not fool ourselves. We realized that in order to be sure, for example, that the shortest program for the IBM 650 that prints the prime numbers has, say, 28 instructions, it would be necessary to prove it, not merely to continue for a long time unsuccessfully trying to discover a program with less than 28 instructions. We could never even sketch a first approach to a proof.

It turns out that it was not our fault that we did not find a proof, because we faced a fundamental limitation. One confronts a very basic difficulty when one tries to prove that a string is random, when one attempts to establish a lower bound on its complexity. We will try to suggest why this problem arises by means of a famous paradox, that of Berry [1, p. 153].

Consider the smallest positive integer that cannot be defined by an English phrase with less than 1 000 000 000 characters. Supposedly the shortest definition of this number has 1 000 000 000 or more characters. However, we defined this number by a phrase much less than 1 000 000 000 characters in length when we described it as "the smallest positive integer that cannot be defined by an English phrase with less than 1 000 000 000 characters!"

What relationship is there between this and proving that a string is complex, that its shortest program needs more than n bits? Consider the first string that can be proven to be of complexity greater than 1 000 000 000. Here once more we face a paradox similar to that of Berry, because this description leads to a program with much less than 1 000 000 000 bits that calculates a string supposedly of complexity greater than 1 000 000 000. Why is there a short program for calculating "the first string that can be proven to be of complexity greater than 1 000 000 000?"

The answer depends on the concept of a formal axiom system, whose importance was emphasized by Hilbert [1]. Hilbert proposed that mathematics be made as exact and precise as possible. In order to avoid arguments between mathematicians about the validity of proofs, he set down explicitly the methods of reasoning used in mathematics. In fact, he invented an artificial language with rules of grammar and spelling that have no exceptions. He proposed that this language be used to eliminate the ambiguities and uncertainties inherent in any natural language. The specifications are so precise and exact that checking if a proof written in this artificial language is correct is completely mechanical. We would say today that it is so clear whether a proof is valid or not that this can be checked by a computer.

Hilbert hoped that this way mathematics would attain the greatest possible objectivity and exactness. Hilbert said that there can no longer be any doubt about proofs. The deductive method should be completely clear.

Suppose that proofs are written in the language that Hilbert constructed, and in accordance with his rules concerning the accepted methods of reasoning. We claim that a computer can be programmed to print all the theorems that can be proven. It is an endless program that every now and then writes on the printer a theorem. Furthermore, no theorem is omitted. Each will eventually be printed, if one is very patient and waits long enough.

How is this possible? The program works in the following manner. The language invented by Hilbert has an alphabet with finitely many signs or characters. First the program generates the strings of characters in this alphabet that are one character in length. It checks if one of these strings satisfies the completely mechanical rules for a correct proof and prints all the theorems whose proofs it has found. Then the program generates all the possible proofs that are two characters in length, and examines each of them to determine if it is valid. The program then examines all possible proofs of length three, of length four, and so on. If a theorem can be proven, the program will eventually find a proof for it in this way, and then print it.

Consider again "the first string that can be proven to be of complexity greater than 1 000 000 000." To find this string one generates all theorems until one finds the first theorem that states that a particular string is of complexity greater than 1 000 000 000. Moreover, the program for finding this string is short, because it need only have the number 1 000 000 000 written in binary notation, $\log_2 1\,000\,000\,000$ bits, and a routine of fixed length c that examines all possible proofs until it finds one that a specific string is of complexity greater than 1 000 000 000.

In fact, we see that there is a program $\log_2 n + c$ bits long that calculates the first string that can be proven to be of complexity greater than n. Here we have Berry's paradox again, because this program of length $\log_2 n + c$ calculates something that supposedly cannot be calculated by a program of length less than or equal to n. Also, $\log_2 n + c$ is much less than n for all sufficiently great values of n, because the logarithm increases very slowly.

What can the meaning of this paradox be? In the case of Berry's original paradox, one cannot arrive at a meaningful conclusion, inasmuch as one is dealing with vague concepts such as an English phrase's defining a positive integer. However our version of the paradox deals with exact concepts that have been defined mathematically. Therefore, it cannot really be a contradiction. It would be absurd for a string not to have a program of length less than or equal to n for calculating it, and at the same time to have such a program. Thus we arrive at the interesting conclusion that such a string cannot exist. For all sufficiently great values of n, one cannot talk about "the first string that can be proven to be of complexity greater than n," because this string cannot exist. In other words, for all sufficiently great values of n, it cannot be proven that a particular string is of complexity greater than n. If one uses the methods of reasoning accepted by Hilbert, there is an upper bound to the complexity that it is possible to prove that a particular string has.

This is the surprising result that we wished to obtain. Most strings of length n are of complexity approximately n, and a string generated by tossing a coin will almost certainly have this property. Nevertheless, one cannot exhibit individual examples of arbitrarily complex strings using methods of reasoning accepted by Hilbert. The lower bounds on the complexity of specific strings that can be established are limited, and we will never be mathematically certain that a particular string is very complex, even though most strings are random.[1]

In 1931 Gödel questioned Hilbert's ideas in a similar way [1], [2]. Hilbert had proposed specifying once and for all exactly what is accepted as a proof, but Gödel explained that no matter what Hilbert specified so precisely, there would always be true statements about the integers that the methods of reasoning accepted by Hilbert would be incapable of proving. This mathematical result has been considered to be of great philosophical importance. Von Neumann commented that the intellectual shock provoked by the crisis in the foundations of mathematics was equaled only by two other scientific events in this century: the theory of relativity and quantum theory [4].

We have combined ideas from information theory and computability theory in order to define the complexity of a binary string, and have then used this concept to give a definition of a random string and to show that a formal axiom system enables one to prove that a random string is indeed random in only finitely many cases.

Now we would like to examine some other possible applications of this

---

[1]This is a particularly perverse example of Kac's comment [13, p. 16] that "as is often the case, it is much easier to prove that an overwhelming majority of objects possess a certain property than to *exhibit* even one such object." The most familiar example of this is Shannon's proof of the coding theorem for a noisy channel; while it is shown that most coding schemes achieve close to the channel capacity, in practice it is difficult to implement a good coding scheme.

viewpoint. In particular, we would like to suggest that the concept of the complexity of a string and the fundamental methodological problems of science are intimately related. We will also suggest that this concept may be of theoretical value in biology.

Solomonoff [7] and the author [9] proposed that the concept of complexity might make it possible to precisely formulate the situation that a scientist faces when he has made observations and wishes to understand them and make predictions. In order to do this the scientist searches for a theory that is in agreement with all his observations. We consider his observations to be represented by a binary string, and a theory to be a program that calculates this string. Scientists consider the simplest theory to be the best one, and that if a theory is too "ad hoc," it is useless. How can we formulate these intuitions about the scientific method in a precise fashion? The simplicity of a theory is inversely proportional to the length of the program that constitutes it. That is to say, the best program for understanding or predicting observations is the shortest one that reproduces what the scientist has observed up to that moment. Also, if the program has the same number of bits as the observations, then it is useless, because it is too "ad hoc." If a string of observations only has theories that are programs with the same length as the string of observations, then the observations are random, and can neither be comprehended nor predicted. They are what they are, and that is all; the scientist cannot have a theory in the proper sense of the concept; he can only show someone else what he observed and say "it was this."

In summary, the value of a scientific theory is that it enables one to compress many observations into a few theoretical hypotheses. There is a theory only when the string of observations is not random, that is to say, when its complexity is appreciably less than its length in bits. In this case the scientist can communicate his observations to a colleague much more economically than by just transmitting the string of observations. He does this by sending his colleague the program that is his theory, and this program must have much fewer bits than the original string of observations.

It is also possible to make a similar analysis of the deductive method, that is to say, of formal axiom systems. This is accomplished by analyzing more carefully the new version of Berry's paradox that was presented. Here we only sketch the three basic results that are obtained in this manner.[2]

1. In a formal system with n bits of axioms it is impossible to prove that

---

[2]See the Appendix.

a particular binary string is of complexity greater than $n + c$.

2. Contrariwise, there are formal systems with $n + c$ bits of axioms in which it is possible to determine each string of complexity less than n and the complexity of each of these strings, and it is also possible to exhibit each string of complexity greater than or equal to n, but without being able to know by how much the complexity of each of these strings exceeds n.

3. Unfortunately, any formal system in which it is possible to determine each string of complexity less than n has either one grave problem or another. Either it has few bits of axioms and needs incredibly long proofs, or it has short proofs but an incredibly great number of bits of axioms. We say "incredibly" because these quantities increase more quickly than any computable function of n.

It is necessary to clarify the relationship between this and the preceding analysis of the scientific method. There are less than 2^n strings of complexity less than n, but some of them are incredibly long. If one wishes to communicate all of them to someone else, there are two alternatives. The first is to directly show all of them to him. In this case one will have to send him an incredibly long message because some of these strings are incredibly long. The other alternative is to send him a very short message consisting of n bits of axioms from which he can deduce which strings are of complexity less than n. Although the message is very short in this case, he will have to spend an incredibly long time to deduce from these axioms the strings of complexity less than n. This is analogous to the dilemma of a scientist who must choose between directly publishing his observations, or publishing a theory that explains them, but requires very extended calculations in order to do this.

Finally, we would like to suggest that the concept of complexity may possibly be of theoretical value in biology.

At the end of his life von Neumann tried to lay the foundation for a mathematics of biological phenomena. His first effort in this direction was his work *Theory of Games and Economic Behavior,* in which he analyzes what is a rational way to behave in situations in which there are conflicting interests [3]. *The Computer and the Brain,* his notes for a lecture series, was published shortly after his death [5]. This book discusses the differences and similarities between the computer and the brain, as a first step to a theory of how the brain functions. A decade later his work *Theory of Self-Reproducing Automata* appeared, in which von Neumann constructs an artificial universe and within it a computer that is capable of reproducing itself [6]. But von Neumann points out that the problem of formulating a mathematical theory of the evolution of life in this abstract setting remains to be solved; and to express mathematically the evolution of the complexity of organisms, one must first define complexity precisely.[3] We submit that "organism" must also be defined, and have tried elsewhere to suggest how this might perhaps be done [10].

We believe that the concept of complexity that has been presented here may be the tool that von Neumann felt is needed. It is by no means accidental that biological phenomena are considered to be extremely complex. Consider how a human being analyzes what he sees, or uses natural languages to communicate. We cannot carry out these tasks by computer because they are as yet too complex for us—the programs would be too long.[4]

Appendix

In this Appendix we try to give a more detailed idea of how the results concerning formal axiom systems that were stated are established.[5]

Two basic mathematical concepts that are employed are the concepts of a recursive function and a partial recursive function. A function is recursive if there is an algorithm for calculating its value when one is given the value of its arguments, in other words, if there is a Turing machine for doing this. If it is possible that this algorithm never terminates and the function is thus undefined for some values of its arguments, then the function is called partial recursive.[6]

In what follows we are concerned with computations involving binary strings. The binary strings are considered to be ordered in the following manner: Λ, 0, 1, 00, 01, 10, 11, 000, 001, 010, ... The natural number n is represented by the nth binary string ($n = 0, 1, 2, \ldots$). The length of a binary

---

[3]In an important paper [14], Eigen studies these questions from the point of view of thermodynamics and biochemistry.

[4]Chandrasekaran and Reeker [15] discuss the relevance of complexity to artificial intelligence.

[5]See [11], [12] for different approaches.

[6]Full treatments of these concepts can be found in standard texts, e.g., Rogers [16].

string s is denoted $\lg(s)$. Thus if s is considered to be a natural number, then $\lg(s) = [\log_2(s+1)]$. Here $[x]$ is the greatest integer $\leq x$.

Definition 1. A *computer* is a partial recursive function $C(p)$. Its argument p is a binary string. The value of $C(p)$ is the binary string output by the computer C when it is given the program p. If $C(p)$ is undefined, this means that running the program p on C produces an unending computation.

Definition 2. The *complexity* $I_C(s)$ of a binary string s is defined to be the length of the shortest program p that makes the computer C output s, i.e.,

$$I_C(s) = \min_{C(p)=s} \lg(p).$$

If no program makes C output s, then $I_C(s)$ is defined to be infinite.

Definition 3. A computer U is *universal* if for any computer C and any binary string s, $I_U(s) \leq I_C(s) + c$, where the constant c depends only on C.

It is easy to see that there are universal computers. For example, consider the computer U such that $U(0^i 1p) = C_i(p)$, where C_i is the ith computer, i.e., a program for U consists of two parts: the left-hand part indicates which computer is to be simulated, and the right-hand part gives the program to be simulated. We now suppose that some particular universal computer U has been chosen as the standard one for measuring complexities, and shall henceforth write $I(s)$ instead of $I_U(s)$.

Definition 4. The *rules of inference* of a class of formal axiom systems is a recursive function $F(a, h)$ (a a binary string, h a natural number) with the property that $F(a, h) \subset F(a, h + 1)$. The value of $F(a, h)$ is the finite (possibly empty) set of theorems that can be proven from the axioms a by means of proofs $\leq h$ characters in length. $F(a) = \bigcup_h F(a, h)$ is the set of theorems that are consequences of the axioms a. The ordered pair $\langle F, a \rangle$, which implies both the choice of rules of inference and axioms, is a particular formal axiom system.

This is a fairly abstract definition, but it retains all those features of formal axiom systems that we need. Note that although one may not be interested in some axioms (e.g., if they are false or incomprehensible), it is stipulated that $F(a, h)$ is always defined.

Theorem 1. a) There is a constant c such that $I(s) \leq \lg(s) + c$ for all binary strings s. b) There are less than 2^n binary strings of complexity less than n.

Proof of a). There is a computer C such that $C(p) = p$ for all programs p. Thus for all binary strings s, $I(s) \leq I_C(s) + c = \lg(s) + c$.

Proof of b). As there are less than 2^n programs of length less than n, there must be less than this number of binary strings of complexity less than n. Q.E.D.

Thesis. A random binary string s is one having the property that $I(s) \approx \lg(s)$.

Theorem 2. Consider the rules of inference F. Suppose that a proposition of the form "$I(s) \geq n$" is in $F(a)$ only if it is true, i.e., only if $I(s) \geq n$. Then a proposition of the form "$I(s) \geq n$" is in $F(a)$ only if $n \leq \lg(a) + c$, where c is a constant that depends only on F.

Proof. Consider that binary string s_k having the shortest proof from the axioms a that it is of complexity $> \lg(a) + 2k$. We claim that $I(s_k) \leq \lg(a) + k + c'$, where c' depends only on F. Taking $k = c'$, we conclude that the binary string $s_{c'}$ with the shortest proof from the axioms a that it is of complexity $> \lg(a) + 2c'$ is, in fact, of complexity $\leq \lg(a) + 2c'$, which is impossible. It follows that s_k doesn't exist for $k = c'$, that is, no binary string can be proven from the axioms a to be of complexity $> \lg(a) + 2c'$. Thus the theorem is proved with $c = 2c'$.

It remains to verify the claim that $I(s_k) \leq \lg(a) + k + c'$. Consider the computer C that does the following when it is given the program $0^k 1a$. It calculates $F(a, h)$ for $h = 0, 1, 2, \ldots$ until it finds the first theorem in $F(a, h)$ of the form "$I(s) \geq n$" with $n > \lg(a) + 2k$. Finally C outputs the binary string s in the theorem it has found. Thus $C(0^k 1a)$ is equal to s_k, if s_k exists. It follows that

$$
\begin{aligned}
I(s_k) &= I(C(0^k 1a)) \\
&\leq I_C(C(0^k 1a)) + c'' \\
&\leq \lg(0^k 1a) + c'' = \lg(a) + k + (c'' + 1) = \lg(a) + k + c'.
\end{aligned}
$$

Q.E.D.

Definition 5. A_n is defined to be the kth binary string of length n, where k is the number of programs p of length $< n$ for which $U(p)$ is defined, i.e., A_n has n and this number k coded into it.

Theorem 3. There are rules of inference F^1 such that for all n, $F^1(A_n)$ is the union of the set of all true propositions of the form "$I(s) = k$" with $k < n$ and the set of all true propositions of the form "$I(s) \geq n$."

Proof. From A_n one knows n and for how many programs p of length $< n$ $U(p)$ is defined. One then simulates in parallel, running each program p of length $< n$ on U until one has determined the value of $U(p)$ for each p of length $< n$ for which $U(p)$ is defined. Knowing the value of $U(p)$ for each p of length $< n$ for which $U(p)$ is defined, one easily determines each string of complexity $< n$ and its complexity. What's more, all other strings must be of complexity $\geq n$. This completes our sketch of how all true propositions of the form "$I(s) = k$" with $k < n$ and of the form "$I(s) \geq n$" can be derived from the axiom A_n. Q.E.D.

Recall that we consider the nth binary string to be the natural number n.

Definition 6. The partial function $B(n)$ is defined to be the biggest natural number of complexity $\leq n$, i.e.,

$$B(n) = \max_{I(k) \leq n} k = \max_{\lg(p) \leq n} U(p).$$

Theorem 4. Let f be a partial recursive function that carries natural numbers into natural numbers. Then $B(n) \geq f(n)$ for all sufficiently great values of n.

Proof. Consider the computer C such that $C(p) = f(p)$ for all p.

$$I(f(n)) \leq I_C(f(n)) + c \leq \lg(n) + c = [\log_2(n+1)] + c < n$$

for all sufficiently great values of n. Thus $B(n) \geq f(n)$ for all sufficiently great values of n. Q.E.D.

Theorem 5. Consider the rules of inference F. Let

$$F_n = \bigcup_a F(a, B(n)),$$

where the union is taken over all binary strings a of length $\leq B(n)$, i.e., F_n is the (finite) set of all theorems that can be deduced by means of proofs with not more than $B(n)$ characters from axioms with not more than $B(n)$ bits. Let s_n be the first binary string s not in any proposition of the form "$I(s) = k$" in F_n. Then $I(s_n) \leq n + c$, where the constant c depends only on F.

Proof. We claim that there is a computer C such that if $U(p) = B(n)$, then $C(p) = s_n$. As, by the definition of B, there is a p_0 of length $\leq n$ such that $U(p_0) = B(n)$, it follows that

$$I(s_n) \leq I_C(s_n) + c = I_C(C(p_0)) + c \leq \lg(p_0) + c \leq n + c,$$

which was to be proved.

It remains to verify the claim that there is a C such that if $U(p) = B(n)$, then $C(p) = s_n$. C works as follows. Given the program p, C first simulates running the program p on U. Once C has determined $U(p)$, it calculates $F(a, U(p))$ for all binary strings a such that $\lg(a) \leq U(p)$, and forms the union of these $2^{U(p)+1} - 1$ different sets of propositions, which is F_n if $U(p) = B(n)$. Finally C outputs the first binary string s not in any proposition of the form "$I(s) = k$" in this set of propositions; s is s_n if $U(p) = B(n)$. Q.E.D.

Theorem 6. Consider the rules of inference F. If $F(a, h)$ includes all true propositions of the form "$I(s) = k$" with $k \leq n + c$, then either $\lg(a) > B(n)$ or $h > B(n)$. Here c is a constant that depends only on F.

Proof. This is an immediate consequence of Theorem 5. Q.E.D.

The following theorem gives an upper bound on the size of the proofs in the formal systems $\langle F^1, A_n \rangle$ that were studied in Theorem 3, and also shows that the lower bound on the size of these proofs that is given by Theorem 6 cannot be essentially improved.

Theorem 7. There is a constant c such that for all n $F^1(A_n, B(n + c))$ includes all true propositions of the form "$I(s) = k$" with $k < n$.

Proof. We claim that there is a computer C such that for all n, $C(A_n) = $ the least natural number h such that $F^1(A_n, h)$ includes all true propositions of the form "$I(s) = k$" with $k < n$. Thus the complexity of this value of h is $\leq \lg(A_n) + c = n + c$, and $B(n + c)$ is $\geq$ this value of h, which was to be proved.

It remains to verify the claim. C works as follows when it is given the program A_n. First, it determines each binary string of complexity $< n$ and its complexity, in the manner described in the proof of Theorem 3. Then it calculates $F^1(A_n, h)$ for $h = 0, 1, 2, \ldots$ until all true propositions of the form "$I(s) = k$" with $k < n$ are included in $F^1(A_n, h)$. The final value of h is then output by C. Q.E.D.

References

[1] J. van Heijenoort, Ed., *From Frege to Gödel: A Source Book in Mathematical Logic, 1879–1931*. Cambridge, Mass.: Harvard Univ. Press, 1967.

[2] M. Davis, Ed., *The Undecidable—Basic Papers on Undecidable Propositions, Unsolvable Problems and Computable Functions*. Hewlett, N.Y.: Raven Press, 1965.

[3] J. von Neumann and O. Morgenstern, *Theory of Games and Economic Behavior*. Princeton, N.J.: Princeton Univ. Press, 1944.

[4] —, "Method in the physical sciences," in *John von Neumann—Collected Works.* New York: Macmillan, 1963, vol. 6, no. 35.

[5] —, *The Computer and the Brain.* New Haven, Conn.: Yale Univ. Press, 1958.

[6] —, *Theory of Self-Reproducing Automata.* Urbana, Ill.: Univ. Illinois Press, 1966. (Edited and completed by A. W. Burks.)

[7] R. J. Solomonoff, "A formal theory of inductive inference," *Inform. Contr.,* vol. 7, pp. 1–22, Mar. 1964; also, pp. 224–254, June 1964.

[8] A. N. Kolmogorov, "Logical basis for information theory and probability theory," *IEEE Trans. Inform. Theory,* vol. IT-14, pp. 662–664, Sept. 1968.

[9] G. J. Chaitin, "On the difficulty of computations," *IEEE Trans. Inform. Theory,* vol. IT-16, pp. 5–9, Jan. 1970.

[10] —, "To a mathematical definition of 'life'," *ACM SICACT News,* no. 4, pp. 12–18, Jan. 1970.

[11] —, "Computational complexity and Gödel's incompleteness theorem," (Abstract) *AMS Notices,* vol. 17, p. 672, June 1970; (Paper) *ACM SIGACT News,* no. 9, pp. 11–12, Apr. 1971.

[12] —, "Information-theoretic limitations of formal systems," presented at the Courant Institute Computational Complexity Symp., N.Y., Oct. 1971. A revised version will appear in *J. Ass. Comput. Mach.*

[13] M. Kac, *Statistical Independence in Probability, Analysis, and Number Theory,* Carus Math. Mono., Mathematical Association of America, no. 12, 1959.

[14] M. Eigen, "Selforganization of matter and the evolution of biological macro-molecules," *Die Naturwissenschaften,* vol. 58, pp. 465–523, Oct. 1971.

[15] B. Chandrasekaran and L. H. Reeker, "Artificial intelligence—a case for agnosticism," Ohio State University, Columbus, Ohio, Rep. OSU-CISRC-TR-72-9, Aug. 1972; also, *IEEE Trans. Syst., Man, Cybern.,* vol. SMC-4, pp. 88–94, Jan. 1974.

[16] H. Rogers, Jr., *Theory of Recursive Functions and Effective Computability.* New York: McGraw-Hill, 1967.

Randomness and mathematical proof

Although randomness can be precisely defined and can even be measured, a given number cannot be proved to be random. This enigma establishes a limit to what is possible in mathematics.

Almost everyone has an intuitive notion of what a random number is. For example, consider these two series of binary digits:

$$01010101010101010101$$
$$01101100110111100010$$

The first is obviously constructed according to a simple rule; it consists of the number 01 repeated ten times. If one were asked to speculate on how the series might continue, one could predict with considerable confidence that the next two digits would be 0 and 1. Inspection of the second series of digits yields no such comprehensive pattern. There is no obvious rule governing the formation of the number, and there is no rational way to guess the succeeding digits. The arrangement seems haphazard; in other words, the sequence appears to be a random assortment of 0's and 1's.

The second series of binary digits was generated by flipping a coin 20 times and writing a 1 if the outcome was heads and a 0 if it was tails. Tossing a coin is a classical procedure for producing a random number, and one might think at first that the provenance of the series alone would certify that it is random. This is not so. Tossing a coin 20 times can produce any one of 2^{20} (or a little more than a million) binary series, and each of them has exactly the same probability. Thus it should be no more surprising to obtain the series with an obvious pattern than to obtain the one that seems to be random; each represents an event with a probability of 2^{-20}. If origin in a probabilistic event were made the sole criterion of randomness, then both series would have to be considered random, and indeed so would all others, since the same mechanism can generate all the possible series. The conclusion is singularly unhelpful in distinguishing the random from the orderly.

Clearly a more sensible definition of randomness is required, one that does not contradict the intuitive concept of a "patternless" number. Such a definition has been devised only in the past 10 years. It does not consider the origin of a number but depends entirely on the characteristics of the sequence of digits. The new definition enables us to describe the properties of a random number more precisely than was formerly possible, and it establishes a hierarchy of degrees of randomness. Of perhaps even greater interest than the capabilities of the definition, however, are its limitations. In particular the definition cannot help to determine, except in very special cases, whether or not a given series of digits, such as the second one above, is in fact random or only seems to be random. This limitation is not a flaw in the definition; it is a consequence of a subtle but fundamental anomaly in the foundation of mathematics. It is closely related to a famous theorem devised and proved in 1931 by Kurt Gödel, which has come to be known as Gödel's incompleteness theorem. Both the theorem and the recent discoveries concerning the nature of randomness help to define the boundaries that constrain certain mathematical methods.

Algorithmic Definition

The new definition of randomness has its heritage in information theory, the science, developed mainly since World War II, that studies the transmission of messages. Suppose you have a friend who is visiting a planet in another galaxy, and that sending him telegrams is very expensive. He forgot to take along his tables of trigonometric functions, and he has asked you to supply them. You could simply translate the numbers into an appropriate code (such as the binary numbers) and transmit them directly, but even the most modest tables of the six functions have a few thousand digits, so that the cost would be high. A much cheaper way to convey the same information would be to transmit instructions for calculating the tables from the underlying trigonometric formulas, such as Euler's equation $e^{ix} = \cos x + i \sin x$. Such a message could be relatively brief, yet inherent in it is all the information contained in even the largest tables.

Suppose, on the other hand, your friend is interested not in trigonometry but in baseball. He would like to know the scores of all the major-league games played since he left the earth some thousands of years before. In this case it is most unlikely that a formula could be found for compressing the information into a short message; in such a series of numbers each digit is essentially an independent item of information, and it cannot be predicted from its neighbors or from some underlying rule. There is no alternative to transmitting the entire list of scores.

In this pair of whimsical messages is the germ of a new definition of randomness. It is based on the observation that the information embodied in a random series of numbers cannot be "compressed," or reduced to a more compact form. In formulating the actual definition it is preferable to consider communication not with a distant friend but with a digital computer. The friend might have the wit to make inferences about numbers or to construct a series from partial information or from vague instructions. The computer does not have that capacity, and for our purposes that deficiency is an advantage. Instructions given the computer must be complete and explicit, and they must enable it to proceed step by step without requiring that it comprehend the result of any part of the operations it performs. Such a program of instructions is an algorithm. It can demand any finite number of mechanical manipulations of numbers, but it cannot ask for judgments about their meaning.

The definition also requires that we be able to measure the information content of a message in some more precise way than by the cost of sending it as a telegram. The fundamental unit of information is the "bit," defined as the smallest item of information capable of indicating a choice between two equally likely things. In binary notation one bit is equivalent to one digit, either a 0 or a 1.

We are now able to describe more precisely the differences between the two series of digits presented at the beginning of this article:

01010101010101010101

01101100110111100010

The first could be specified to a computer by a very simple algorithm, such as "Print 01 ten times." If the series were extended according to the same rule, the algorithm would have to be only slightly larger; it might be made to read, for example, "Print 01 a million times." The number of bits in such an algorithm is a small fraction of the number of bits in the series it specifies, and as the series grows larger the size of the program increases at a much slower rate.

For the second series of digits there is no corresponding shortcut. The most economical way to express the series is to write it out in full, and the shortest algorithm for introducing the series into a computer would be "Print 01101100110111100010." If the series were much larger (but still apparently patternless), the algorithm would have to be expanded to the corresponding size. This "incompressibility" is a property of all random numbers; indeed, we can proceed directly to define randomness in terms of incompressibility: A series of numbers is random if the smallest algorithm capable of specifying it to a computer has about the same number of bits of information as the series itself.

This definition was independently proposed about 1965 by A. N. Kolmogorov of the Academy of Science of the U.S.S.R. and by me, when I was an undergraduate at the City College of the City University of New York. Both Kolmogorov and I were then unaware of related proposals made in 1960 by Ray J. Solomonoff of the Zator Company in an endeavor to measure the simplicity of scientific theories. During the past decade we and others have continued to explore the meaning of randomness. The original formulations have been improved and the feasibility of the approach has been amply confirmed.

Model of Inductive Method

The algorithmic definition of randomness provides a new foundation for the theory of probability. By no means does it supersede classical probability theory, which is based on an ensemble of possibilities, each of which is assigned a probability. Rather, the algorithmic approach complements the ensemble method by giving precise meaning to concepts that had been intuitively appealing but that could not be formally adopted.

The ensemble theory of probability, which originated in the 17th century, remains today of great practical importance. It is the foundation of statistics, and it is applied to a wide range of problems in science and engineering. The algorithmic theory also has important implications, but they are primarily theoretical. The area of broadest interest is its amplification of Gödel's incompleteness theorem. Another application (which actually preceded the formulation of the theory itself) is in Solomonoff's model of scientific induction.

Solomonoff represented a scientist's observations as a series of binary digits. The scientist seeks to explain these observations through a theory, which can be regarded as an algorithm capable of generating the series and extending it, that is, predicting future observations. For any given series of observations there are always several competing theories, and the scientist must choose among them. The model demands that the smallest algorithm, the one consisting of the fewest bits, be selected. Stated another way, this rule is the familiar formulation of Occam's razor: Given differing theories of apparently equal merit, the simplest is to be preferred.

Thus in the Solomonoff model a theory that enables one to understand a series of observations is seen as a small computer program that reproduces the observations and makes predictions about possible future observations. The smaller the program, the more comprehensive the theory and the greater the degree of understanding. Observations that are random cannot be reproduced by a small program and therefore cannot be explained by a theory. In addition the future behavior of a random system cannot be predicted. For random data the most compact way for the scientist to communicate his observations is for him to publish them in their entirety.

Defining randomness or the simplicity of theories through the capabilities of the digital computer would seem to introduce a spurious element into these essentially abstract notions: the peculiarities of the particular computing machine employed. Different machines communicate through different computer languages, and a set of instructions expressed in one of those languages might require more or fewer bits when the instructions are translated into another language. Actually, however, the choice of computer matters very little. The problem can be avoided entirely simply by insisting that the randomness of all numbers be tested on the same machine. Even when different machines are employed, the idiosyncrasies of various languages can readily be compensated for. Suppose, for example, someone has a program written in English and wishes to utilize it with a computer that reads only French. Instead of translating the algorithm itself he could preface the program with a complete English course written in French. Another mathematician with a French program and an English machine would follow the opposite procedure. In this way only a fixed number of bits need be added to the program, and that number grows less significant as the size of the series specified by the program increases. In practice a device called a compiler often makes it possible to ignore the differences between languages when one is addressing a computer.

Since the choice of a particular machine is largely irrelevant, we can choose for our calculations an ideal computer. It is assumed to have unlimited storage capacity and unlimited time to complete its calculations. Input to and output from the machine are both in the form of binary digits. The machine begins to operate as soon as the program is given it, and it continues until it has finished printing the binary series that is the result. The machine then halts. Unless an error is made in the program, the computer will produce exactly one output for any given program.

Minimal Programs and Complexity

Any specified series of numbers can be generated by an infinite number of algorithms. Consider, for example, the three-digit decimal series 123. It could be produced by an algorithm such as "Subtract 1 from 124 and print the result," or "Subtract 2 from 125 and print the result," or an infinity of other programs formed on the same model. The programs of greatest interest, however, are the smallest ones that will yield a given numerical series. The smallest programs are called minimal programs; for a given series there may be only one minimal program or there may be many.

Any minimal program is necessarily random, whether or not the series it generates is random. This conclusion is a direct result of the way we have defined randomness. Consider the program P, which is a minimal program for the series of digits S. If we assume that P is not random, then by definition there must be another program, P', substantially smaller than P that will generate it. We can then produce S by the following algorithm: "From P' calculate P, then from P calculate S." This program is only a few bits longer than P', and thus it must be substantially shorter than P. P is therefore not a minimal program.

The minimal program is closely related to another fundamental concept in the algorithmic theory of randomness: the concept of complexity. The complexity of a series of digits is the number of bits that must be put into a computing machine in order to obtain the original series as output. The complexity is therefore equal to the size in bits of the minimal programs of the series. Having introduced this concept, we can now restate our definition of randomness in more rigorous terms: A random series of digits is one whose complexity is approximately equal to its size in bits.

The notion of complexity serves not only to define randomness but also to measure it. Given several series of numbers each having n digits, it is theoretically possible to identify all those of complexity $n-1$, $n-10$, $n-100$ and so forth and thereby to rank the series in decreasing order of randomness. The exact value of complexity below which a series is no longer considered random remains somewhat arbitrary. The value ought to be set low enough for numbers with obviously random properties not to be excluded and high enough for numbers with a conspicuous pattern to be disqualified, but to set a particular numerical value is to judge what degree of randomness constitutes actual randomness. It is this uncertainty that is reflected in the qualified statement that the complexity of a random series is *approximately* equal to the size of the series.

Properties of Random Numbers

The methods of the algorithmic theory of probability can illuminate many of the properties of both random and nonrandom numbers. The frequency distribution of digits in a series, for example, can be shown to have an important influence on the randomness of the series. Simple inspection suggests that a series consisting entirely of either 0's or 1's is far from random, and the algorithmic approach confirms that conclusion. If such a series is n digits long, its complexity is approximately equal to the logarithm to the base 2 of n. (The exact value depends on the machine language employed.) The series can be produced by a simple algorithm such as "Print 0 n times," in which virtually all the information needed is contained in the binary numeral for n. The size of this number is about $\log_2 n$ bits. Since for even a moderately long series the logarithm of n is much smaller than n itself, such numbers are of low complexity; their intuitively perceived pattern is mathematically confirmed.

Another binary series that can be profitably analyzed in this way is one where 0's and 1's are present with relative frequencies of three-fourths and one-fourth. If the series is of size n, it can be demonstrated that its complexity is no greater than four-fifths n, that is, a program that will produce the series can be written in $4n/5$ bits. This maximum applies regardless of the sequence of the digits, so that no series with such a frequency distribution can be considered very random. In fact, it can be proved that in any long binary series that is random the relative frequencies of 0's and 1's must be very close to one-half. (In a random decimal series the relative frequency of each digit is, of course, one-tenth.)

Numbers having a nonrandom frequency distribution are exceptional. Of all the possible n-digit binary numbers there is only one, for example, that consists entirely of 0's and only one that is all 1's. All the rest are less orderly, and the great majority must, by any reasonable standard, be called random. To choose an arbitrary limit, we can calculate the fraction of all n-digit binary numbers that have a complexity of less than $n - 10$. There are 2^1 programs one digit long that might generate an n-digit series; there are 2^2 programs two digits long that could yield such a series, 2^3 programs three digits long and so forth, up to the longest programs permitted within the allowed complexity; of these there are 2^{n-11}. The sum of this series $(2^1 + 2^2 + \cdots + 2^{n-11})$ is equal to $2^{n-10} - 2$. Hence there are fewer than 2^{n-10} programs of size less than $n-10$, and since each of these programs can specify no more than one series of digits, fewer than 2^{n-10} of the 2^n numbers have a complexity less than $n - 10$. Since $2^{n-10}/2^n = 1/1,024$, it follows that of all the n-digit binary numbers only about one in 1,000 have a complexity less than $n-10$. In other words, only about one series in 1,000 can be compressed into a computer program more than 10 digits smaller than itself.

A necessary corollary of this calculation is that more than 999 of every 1,000 n-digit binary numbers have a complexity equal to or greater than $n - 10$. If that degree of complexity can be taken as an appropriate test of randomness, then almost all n-digit numbers are in fact random. If a fair coin is tossed n times, the probability is greater than .999 that the result will be random to this extent. It would therefore seem easy to exhibit a specimen of a long series of random digits; actually it is impossible to do so.

Formal Systems

It can readily be shown that a specific series of digits is not random; it is sufficient to find a program that will generate the series and that is substantially smaller than the series itself. The program need not be a minimal program for the series; it need only be a small one. To demonstrate that a particular series of digits is random, on the other hand, one must prove that no small program for calculating it exists.

It is in the realm of mathematical proof that Gödel's incompleteness theorem is such a conspicuous landmark; my version of the theorem predicts that the required proof of randomness cannot be found. The consequences of this fact are just as interesting for what they reveal about Gödel's theorem as they are for what they indicate about the nature of random numbers.

Gödel's theorem represents the resolution of a controversy that preoccupied mathematicians during the early years of the 20th century. The question at issue was: "What constitutes a valid proof in mathematics and how is such a proof to be recognized?" David Hilbert had attempted to resolve the controversy by devising an artificial language in which valid proofs could be found mechanically, without any need for human insight or judgement. Gödel showed that there is no such perfect language.

Hilbert established a finite alphabet of symbols, an unambiguous grammar specifying how a meaningful statement could be formed, a finite list of axioms, or initial assumptions, and a finite list of rules of inference for deducing theorems from the axioms or from other theorems. Such a language, with its rules, is called a formal system.

A formal system is defined so precisely that a proof can be evaluated by a recursive procedure involving only simple logical and arithmetical manipulations. In other words, in the formal system there is an algorithm for testing the validity of proofs. Today, although not in Hilbert's time, the algorithm could be executed on a digital computer and the machine could be asked to "judge" the merits of the proof.

Because of Hilbert's requirement that a formal system have a proof-checking algorithm, it is possible in theory to list one by one all the theorems that can be proved in a particular system. One first lists in alphabetical order all sequences of symbols one character long and applies the proof-testing algorithm to each of them, thereby finding all theorems (if any) whose proofs consist of a single character. One then tests all the two-character sequences of symbols, and so on. In this way all potential proofs can be checked, and eventually all theorems can be discovered in order of the size of their proofs. (The method is, of course, only a theoretical one; the procedure is too lengthy to be practical.)

Unprovable Statements

Gödel showed in his 1931 proof that Hilbert's plan for a completely systematic mathematics cannot be fulfilled. He did this by constructing an assertion about the positive integers in the language of the formal system that is true but that cannot be proved in the system. The formal system, no matter how large or how carefully constructed it is, cannot encompass all true theorems and is therefore incomplete. Gödel's technique can be applied to virtually any formal system, and it therefore demands the surprising and, for many, discomforting conclusion that there can be no definitive answer to the question "What is a valid proof?"

Gödel's proof of the incompleteness theorem is based on the paradox of Epimenides the Cretan, who is said to have averred, "All Cretans are liars" [see "Paradox," by W. V. Quine; *Scientific American,* April, 1962]. The paradox can be rephrased in more general terms as "This statement is false," an assertion that is true if and only if it is false and that is therefore neither true nor false. Gödel replaced the concept of truth with that of provability and thereby constructed the sentence "This statement is unprovable," an assertion that, in a specific formal system, is provable if and only if it is false. Thus either a falsehood is provable, which is forbidden, or a true statement is unprovable, and hence the formal system is incomplete. Gödel then applied a technique that uniquely numbers all statements and proofs in the formal system and thereby converted the sentence "This statement is unprovable" into an assertion about the properties of the positive integers. Because this transformation is possible, the incompleteness theorem applies with equal cogency to all formal systems in which it is possible to deal with the positive integers [see "Gödel's Proof," by Ernest Nagel and James R. Newman; *Scientific American,* June, 1956].

The intimate association between Gödel's proof and the theory of random numbers can be made plain through another paradox, similar in form to the paradox of Epimenides. It is a variant of the Berry paradox, first published in 1908 by Bertrand Russell. It reads: "Find the smallest positive integer which to be specified requires more characters than there are in this sentence." The sentence has 114 characters (counting spaces between words and the period but not the quotation marks), yet it supposedly specifies an integer that, by definition, requires more than 114 characters to be specified.

As before, in order to apply the paradox to the incompleteness theorem it is necessary to remove it from the realm of truth to the realm of provability. The phrase "which requires" must be replaced by "which can be proved to require," it being understood that all statements will be expressed in a particular formal system. In addition the vague notion of "the number of characters required to specify" an integer can be replaced by the precisely defined concept of complexity, which is measured in bits rather than characters.

The result of these transformations is the following computer program: "Find a series of binary digits that can be proved to be of a complexity greater than the number of bits in this program." The program tests all possible proofs in the formal system in order of their size until it encounters the first one proving that a specific binary sequence is of a complexity greater than the number of bits in the program. Then it prints the series it has found and halts. Of course, the paradox in the statement from which the program was derived has not been eliminated. The program supposedly calculates a number that no program its size should be able to calculate. In fact, the program finds the first number that it can be proved incapable of finding.

The absurdity of this conclusion merely demonstrates that the program will never find the number it is designed to look for. In a formal system one cannot prove that a particular series of digits is of a complexity greater than the number of bits in the program employed to specify the series.

A further generalization can be made about this paradox. It is not the number of bits in the program itself that is the limiting factor but the number of bits in the formal system as a whole. Hidden in the program are the axioms and rules of inference that determine the behavior of the system and provide the algorithm for testing proofs. The information content of these axioms and rules can be measured and can be designated the complexity of the formal system. The size of the entire program therefore exceeds the complexity of the formal system by a fixed number of bits c. (The actual value of c depends on the machine language employed.) The theorem proved by the paradox can therefore be stated as follows: In a formal system of complexity n it is impossible to prove that a particular series of binary digits is of complexity greater than $n + c$, where c is a constant that is independent of the particular system employed.

Limits of Formal Systems

Since complexity has been defined as a measure of randomness, this theorem implies that in a formal system no number can be proved to be random unless the complexity of the number is less than that of the system itself. Because all minimal programs are random the theorem also implies that a system of greater complexity is required in order to prove that a program is a minimal one for a particular series of digits.

The complexity of the formal system has such an important bearing on the proof of randomness because it is a measure of the amount of information the system contains, and hence of the amount of information that can be derived from it. The formal system rests on axioms: fundamental statements that are irreducible in the same sense that a minimal program is. (If an axiom could be expressed more compactly, then the briefer statement would become a new axiom and the old one would become a derived theorem.) The information embodied in the axioms is thus itself random, and it can be employed to test the randomness of other data. The randomness of some numbers can therefore be proved, but only if they are smaller than the formal system. Moreover, any formal system is of necessity finite, whereas any series of digits can be made arbitrarily large. Hence there will always be numbers whose randomness cannot be proved.

The endeavor to define and measure randomness has greatly clarified the significance and the implications of Gödel's incompleteness theorem. That theorem can now be seen not as an isolated paradox but as a natural consequence of the constraints imposed by information theory. In 1946 Hermann Weyl said that the doubt induced by such discoveries as Gödel's theorem had been "a constant drain on the enthusiasm and determination with which I pursued my research work." From the point of view of information theory, however, Gödel's theorem does not appear to give cause for depression. Instead it seems simply to suggest that in order to progress, mathematicians, like investigators in other sciences, must search for new axioms.

Illustrations

Algorithmic definition of randomness

(a) 10100→Computer→11111111111111111111

(b) 01101100110111100010→Computer→01101100110111100010

Algorithmic definition of randomness relies on the capabilities and limitations of the digital computer. In order to produce a particular output, such as a series of binary digits, the computer must be given a set of explicit instructions that can be followed without making intellectual judgments. Such a program of instructions is an algorithm. If the desired output is highly ordered (a), a relatively small algorithm will suffice; a series of twenty 1's, for example, might be generated by some hypothetical computer from the program 10100, which is the binary notation for the decimal number 20. For a random series of digits (b) the most concise program possible consists of the series itself. The smallest programs capable of generating a particular program are called the minimal programs of the series; the size of these programs, measured in bits, or binary digits, is the complexity of the series. A series of digits is defined as random if series' complexity approaches its size in bits.

Formal systems

Alphabet, Grammar, Axioms, Rules of Inference

↓

Computer

↓

Theorem 1, Theorem 2, Theorem 3, Theorem 4, Theorem 5, ...

Formal systems devised by David Hilbert contain an algorithm that mechanically checks the validity of all proofs that can be formulated in the system. The formal system consists of an alphabet of symbols in which all statements can be written; a grammar that specifies how the symbols are to be combined; a set of axioms, or principles accepted without proof; and rules of inference for deriving theorems from the axioms. Theorems are found by writing all the possible grammatical statements in the system and testing them to determine which ones are in accord with the rules of inference and are therefore valid proofs. Since this operation can be performed by an algorithm it could be done by a digital computer. In 1931 Kurt Gödel demonstrated that virtually all formal systems are incomplete: in each of them there is at least one statement that is true but that cannot be proved.

Inductive reasoning

Observations: 0101010101

Predictions: 01010101010101010101
Theory: Ten repetitions of 01
Size of Theory: 21 characters

Predictions: 01010101010000000000
Theory: Five repetitions of 01 followed by ten 0's
Size of Theory: 42 characters

Inductive reasoning as it is employed in science was analyzed mathematically by Ray J. Solomonoff. He represented a scientist's observations as a series of binary digits; the observations are to be explained and new ones are to be predicted by theories, which are regarded as algorithms instructing a computer to reproduce the observations. (The programs would not be English sentences but binary series, and their size would be measured not in characters but in bits.) Here two competing theories explain the existing data; Occam's razor demands that the simpler, or smaller, theory be preferred. The task of the scientist is to search for minimal programs. If the data are random, the minimal programs are no more concise than the observations and no theory can be formulated.

Random sequences

Illustration is a graph of number of n-digit sequences
as a function of their complexity.

The curve grows exponentially
from approximately 0 to approximately 2^n
as the complexity goes from 0 to n.

Random sequences of binary digits make up the majority of all such sequences. Of the 2^n series of n digits, most are of a complexity that is within a few bits of n. As complexity decreases, the number of series diminishes in a roughly exponential manner. Orderly series are rare; there is only one, for example, that consists of n 1's.

Three paradoxes

RUSSELL PARADOX
Consider the set of all sets that are not members of themselves.
Is this set a member of itself?

EPIMENIDES PARADOX
Consider this statement: "This statement is false."
Is this statement true?

BERRY PARADOX
Consider this sentence: "Find the smallest positive integer
which to be specified requires more characters
than there are in this sentence."
Does this sentence specify a positive integer?

Three paradoxes delimit what can be proved. The first, devised by Bertrand Russell, indicated that informal reasoning in mathematics can yield contradictions, and it led to the creation of formal systems. The second, attributed to Epimenides, was adapted by Gödel to show that even within a formal system there are true statements that are unprovable. The third leads to the demonstration that a specific number cannot be proved random.

Unprovable statements

(a) This statement is unprovable.

(b) The complexity of 01101100110111100010 is greater than 15 bits.

(c) The series of digits 01101100110111100010 is random.

(d) 10100 is a minimal program for the series 11111111111111111111.

Unprovable statements can be shown to be false, if they are false, but they cannot be shown to be true. A proof that "This statement is unprovable" (a) reveals a self-contradiction in a formal system. The assignment of a numerical value to the complexity of a particular number (b) requires a proof that no smaller algorithm for generating the number exists; the proof could be supplied only if the formal system itself were more complex than the number. Statements labeled c and d are subject to the same limitation, since the identification of a random number or a minimal program requires the determination of complexity.

Further Reading

- *A Profile of Mathematical Logic.* Howard DeLong. Addison-Wesley, 1970.

- *Theories of Probability: An Examination of Foundations.* Terrence L. Fine. Academic Press, 1973.

- *Universal Gambling Schemes and the Complexity Measures of Kolmogorov and Chaitin.* Thomas M. Cover. Technical Report No. 12, Statistics Department, Stanford University, 1974.

- "Information-Theoretic Limitations of Formal Systems." Gregory J. Chaitin in *Journal of the Association for Computing Machinery,* Vol. 21, pages 403–424; July, 1974.

Gödel's theorem and information

Gödel's theorem may be demonstrated using arguments having an information-theoretic flavor. In such an approach it is possible to argue that if a theorem contains more information than a given set of axioms, then it is impossible for the theorem to be derived from the axioms. In contrast with the traditional proof based on the paradox of the liar, this new viewpoint suggests that the incompleteness phenomenon discovered by Gödel is natural and widespread rather than pathological and unusual.

1. Introduction

To set the stage, let us listen to Hermann Weyl (1946), as quoted by Eric Temple Bell (1951):

> We are less certain than ever about the ultimate foundations of (logic and) mathematics. Like everybody and everything in the world today, we have our "crisis." We have had it for nearly fifty years. Outwardly it does not seem to hamper our daily work, and yet I for one confess that it has had a considerable practical influence on my mathematical life: it directed my interests to fields I considered relatively "safe," and has been a constant drain on the enthusiasm and determination with which I pursued my research work. This experience is probably shared by other mathematicians who are not indifferent to what their scientific endeavors mean in the context of man's whole caring and knowing, suffering and creative existence in the world.

And these are the words of John von Neumann (1963):

> ...there have been within the experience of people now living
> at least three serious crises... There have been two such crises in
> physics—namely, the conceptual soul-searching connected with
> the discovery of relativity and the conceptual difficulties con-
> nected with discoveries in quantum theory... The third crisis was
> in mathematics. It was a very serious conceptual crisis, dealing
> with rigor and the proper way to carry out a correct mathemat-
> ical proof. In view of the earlier notions of the absolute rigor of
> mathematics, it is surprising that such a thing could have hap-
> pened, and even more surprising that it could have happened in
> these latter days when miracles are not supposed to take place.
> Yet it did happen.

At the time of its discovery, Kurt Gödel's incompleteness theorem was
a great shock and caused much uncertainty and depression among mathematicians sensitive to foundational issues, since it seemed to pull the rug out
from under mathematical certainty, objectivity, and rigor. Also, its proof was
considered to be extremely difficult and recondite. With the passage of time
the situation has been reversed. A great many different proofs of Gödel's
theorem are now known, and the result is now considered easy to prove and
almost obvious: It is equivalent to the unsolvability of the halting problem,
or alternatively to the assertion that there is an r.e. (recursively enumerable)
set that is not recursive. And it has had no lasting impact on the daily lives
of mathematicians or on their working habits; no one loses sleep over it any
more.

Gödel's original proof constructed a paradoxical assertion that is true but
not provable within the usual formalizations of number theory. In contrast I
would like to measure the power of a set of axioms and rules of inference. I
would like to able to say that if one has ten pounds of axioms and a twenty-
pound theorem, then that theorem cannot be derived from those axioms.
And I will argue that this approach to Gödel's theorem does suggest a change
in the daily habits of mathematicians, and that Gödel's theorem cannot be
shrugged away.

To be more specific, I will apply the viewpoint of thermodynamics and
statistical mechanics to Gödel's theorem, and will use such concepts as probability, randomness, entropy, and information to study the incompleteness
phenomenon and to attempt to evaluate how widespread it is. On the basis of this analysis, I will suggest that mathematics is perhaps more akin to

physics than mathematicians have been willing to admit, and that perhaps a more flexible attitude with respect to adopting new axioms and methods of reasoning is the proper response to Gödel's theorem. Probabilistic proofs of primality via sampling (Chaitin and Schwartz, 1978) also suggest that the sources of mathematical truth are wider than usually thought. Perhaps number theory should be pursued more openly in the spirit of experimental science (Pólya, 1959)!

I am indebted to John McCarthy and especially to Jacob Schwartz for making me realize that Gödel's theorem is not an obstacle to a practical AI (artificial intelligence) system based on formal logic. Such an AI would take the form of an intelligent proof checker. Gottfried Wilhelm Liebnitz and David Hilbert's dream that disputes could be settled with the words "Gentlemen, let us compute!" and that mathematics could be formalized, should still be a topic for active research. Even though mathematicians and logicians have erroneously dropped this train of thought dissuaded by Gödel's theorem, great advances have in fact been made "covertly," under the banner of computer science, LISP, and AI (Cole et al., 1981; Dewar et al., 1981; Levin, 1974; Wilf, 1982).

To speak in metaphors from Douglas Hofstadter (1979), we shall now stroll through an art gallery of proofs of Gödel's theorem, to the tune of Moussorgsky's pictures at an exhibition! Let us start with some traditional proofs (Davis, 1978; Hofstadter, 1979; Levin, 1974; Post, 1965).

2. Traditional Proofs of Gödel's Theorem

Gödel's original proof of the incompleteness theorem is based on the paradox of the liar: "This statement is false." He obtains a theorem instead of a paradox by changing this to: "This statement is unprovable." If this assertion is unprovable, then it is true, and the formalization of number theory in question is incomplete. If this assertion is provable, then it is false, and the formalization of number theory is inconsistent. The original proof was quite intricate, much like a long program in machine language. The famous technique of Gödel numbering statements was but one of the many ingenious ideas brought to bear by Gödel to construct a number-theoretic assertion which says of itself that it is unprovable.

Gödel's original proof applies to a particular formalization of number theory, and was to be followed by a paper showing that the same methods applied to a much broader class of formal axiomatic systems. The modern approach in fact applies to all formal axiomatic systems, a concept which could not even be defined when Gödel wrote his original paper, owing to the lack of a mathematical definition of effective procedure or computer algorithm. After Alan Turing succeeded in defining effective procedure by inventing a simple idealized computer now called the Turing machine (also done independently by Emil Post), it became possible to proceed in a more general fashion.

Hilbert's key requirement for a formal mathematical system was that there be an objective criterion for deciding if a proof written in the language of the system is valid or not. In other words, there must be an algorithm, a computer program, a Turing machine, for checking proofs. And the compact modern definition of formal axiomatic system as a recursively enumerable set of assertions is an immediate consequence if one uses the so-called British Museum algorithm. One applies the proof checker in turn to all possible proofs, and prints all the theorems, which of course would actually take astronomical amounts of time. By the way, in practice LISP is a very convenient programming language in which to write a simple proof checker (Levin, 1974).

Turing showed that the halting problem is unsolvable, that is, that there is no effective procedure or algorithm for deciding whether or not a program ever halts. Armed with the general definition of a formal axiomatic system as an r.e. set of assertions in a formal language, one can immediately deduce a version of Gödel's incompleteness theorem from Turing's theorem. I will sketch three different proofs of the unsolvability of the halting problem in a moment; first let me derive Gödel's theorem from it. The reasoning is simply that if it were always possible to prove whether or not particular programs halt, since the set of theorems is r.e., one could use this to solve the halting problem for any particular program by enumerating all theorems until the matter is settled. But this contradicts the unsolvability of the halting problem.

Here come three proofs that the halting problem is unsolvable. One proof considers that function $F(N)$ defined to be either one more than the value of the Nth computable function applied to the natural number N, or zero if this value is undefined because the Nth computer program does not halt on input N. F cannot be a computable function, for if program N calculated it, then one would have $F(N) = F(N) + 1$, which is impossible. But the only way that F can fail to be computable is because one cannot decide if the Nth program ever halts when given input N.

The proof I have just given is of course a variant of the diagonal method which Georg Cantor used to show that the real numbers are more numerous than the natural numbers (Courant and Robbins, 1941). Something much closer to Cantor's original technique can also be used to prove Turing's theorem. The argument runs along the lines of Bertrand Russell's paradox (Russell, 1967) of the set of all things that are not members of themselves. Consider programs for enumerating sets of natural numbers, and number these computer programs. Define a set of natural numbers consisting of the numbers of all programs which do not include their own number in their output set. This set of natural numbers cannot be recursively enumerable, for if it were listed by computer program N, one arrives at Russell's paradox of the barber in a small town who shaves all those and only those who do not shave themselves, and can neither shave himself nor avoid doing so. But the only way that this set can fail to be recursively enumerable is if it is impossible to decide whether or not a program ever outputs a specific natural number, and this is a variant of the halting problem.

For yet another proof of the unsolvability of the halting problem, consider programs which take no input and which either produce a single natural number as output or loop forever without ever producing an output. Think of these programs as being written in binary notation, instead of as natural numbers as before. I now define a so-called Busy Beaver function: BB of N is the largest natural number output by any program less than N bits in size. The original Busy Beaver function measured program size in terms of the number of states in a Turing machine instead of using the more correct information-theoretic measure, bits. It is easy to see that BB of N grows more quickly than any computable function, and is therefore not computable, which as before implies that the halting problem is unsolvable.

In a beautiful and easy to understand paper Post (1965) gave versions of Gödel's theorem based on his concepts of simple and creative r.e. sets. And he formulated the modern abstract form of Gödel's theorem, which is like a Japanese haiku: there is an r.e. set of natural numbers that is not recursive. This set has the property that there are programs for printing all the members of the set in some order, but not in ascending order. One can eventually realize that a natural number is a member of the set, but there is no algorithm for deciding if a given number is in the set or not. The set is r.e. but its complement is not. In fact, the set of (numbers of) halting programs is such a set. Now consider a particular formal axiomatic system in which one can talk about natural numbers and computer programs and such, and let X be any r.e. set whose complement is not r.e. It follows immediately that not all true assertions of the form "the natural number N is not a member of the set X" are theorems in the formal axiomatic system. In fact, if X is what Post called a simple r.e. set, then only finitely many of these assertions can be theorems.

These traditional proofs of Gödel's incompleteness theorem show that formal axiomatic systems are incomplete, but they do not suggest ways to measure the power of formal axiomatic systems, to rank their degree of completeness or incompleteness. Actually, Post's concept of a simple set contains the germ of the information-theoretic versions of Gödel's theorem that I will give later, but this is only visible in retrospect. One could somehow choose a particular simple r.e. set X and rank formal axiomatic systems according to how many different theorems of the form "N is not in X" are provable. Here are three other quantitative versions of Gödel's incompleteness theorem which do sort of fall within the scope of traditional methods.

Consider a particular formal axiomatic system in which it is possible to talk about total recursive functions (computable functions which have a natural number as value for each natural number input) and their running time computational complexity. It is possible to construct a total recursive function which grows more quickly than any function which is provably total recursive in the formal axiomatic system. It is also possible to construct a total recursive function which takes longer to compute than any provably total recursive function. That is to say, a computer program which produces a natural number output and then halts whenever it is given a natural number input, but this cannot be proved in the formal axiomatic system, because the program takes too long to produce its output.

It is also fun to use constructive transfinite ordinal numbers (Hofstadter, 1979) to measure the power of formal axiomatic systems. A constructive ordinal is one which can be obtained as the limit from below of a computable sequence of smaller constructive ordinals. One measures the power of a formal axiomatic system by the first constructive ordinal which cannot be proved to be a constructive ordinal within the system. This is like the paradox of the first unmentionable or indefinable ordinal number (Russell, 1967)!

Before turning to information-theoretic incompleteness theorems, I must first explain the basic concepts of algorithmic information theory (Chaitin, 1975b, 1977, 1982).

3. Algorithmic Information Theory

Algorithmic information theory focuses on individual objects rather than on the ensembles and probability distributions considered in Claude Shannon and Norbert Wiener's information theory. How many bits does it take to describe how to compute an individual object? In other words, what is the size in bits of the smallest program for calculating it? It is easy to see that since general-purpose computers (universal Turing machines) can simulate each other, the choice of computer as yardstick is not very important and really only corresponds to the choice of origin in a coordinate system.

The fundamental concepts of this new information theory are: algorithmic information content, joint information, relative information, mutual information, algorithmic randomness, and algorithmic independence. These are defined roughly as follows.

The algorithmic information content $I(X)$ of an individual object X is defined to be the size of the smallest program to calculate X. Programs must be self-delimiting so that subroutines can be combined by concatenating them. The joint information $I(X, Y)$ of two objects X and Y is defined to be the size of the smallest program to calculate X and Y simultaneously. The relative or conditional information content $I(X|Y)$ of X given Y is defined to be the size of the smallest program to calculate X from a minimal program for Y.

Note that the relative information content of an object is never greater than its absolute information content, for being given additional information can only help. Also, since subroutines can be concatenated, it follows that joint information is subadditive. That is to say, the joint information content is bounded from above by the sum of the individual information contents of the objects in question. The extent to which the joint information content is less than this sum leads to the next fundamental concept, mutual information.

The mutual information content $I(X : Y)$ measures the commonality of X and Y: it is defined as the extent to which knowing X helps one to calculate Y, which is essentially the same as the extent to which knowing Y helps one to calculate X, which is also the same as the extent to which it is cheaper to calculate them together than separately. That is to say,

$$
\begin{aligned}
I(X : Y) &= I(X) - I(X|Y) \\
&= I(Y) - I(Y|X)
\end{aligned}
$$

$$= I(X) + I(Y) - I(X,Y).$$

Note that this implies that

$$\begin{aligned} I(X,Y) &= I(X) + I(Y|X) \\ &= I(Y) + I(X|Y). \end{aligned}$$

I can now define two very fundamental and philosophically significant notions: algorithmic randomness and algorithmic independence. These concepts are, I believe, quite close to the intuitive notions that go by the same name, namely, that an object is chaotic, typical, unnoteworthy, without structure, pattern, or distinguishing features, and is irreducible information, and that two objects have nothing in common and are unrelated.

Consider, for example, the set of all N-bit long strings. Most such strings S have $I(S)$ approximately equal to N plus $I(N)$, which is N plus the algorithmic information contained in the base-two numeral for N, which is equal to N plus order of $\log N$. No N-bit long S has information content greater than this. A few have less information content; these are strings with a regular structure or pattern. Those S of a given size having greatest information content are said to be random or patternless or algorithmically incompressible. The cutoff between random and nonrandom is somewhere around $I(S)$ equal to N if the string S is N bits long.

Similarly, an infinite binary sequence such as the base-two expansion of π is random if and only if all its initial segments are random, that is, if and only if there is a constant C such that no initial segment has information content less than C bits below its length. Of course, π is the extreme opposite of a random string: it takes only $I(N)$ which is order of $\log N$ bits to calculate π's first N bits. But the probability that an infinite sequence obtained by independent tosses of a fair coin is algorithmically random is unity.

Two strings are algorithmically independent if their mutual information is essentially zero, more precisely, if their mutual information is as small as possible. Consider, for example, two arbitrary strings X and Y each N bits in size. Usually, X and Y will be random to each other, excepting the fact that they have the same length, so that $I(X : Y)$ is approximately equal to $I(N)$. In other words, knowing one of them is no help in calculating the other, excepting that it tells one the other string's size.

To illustrate these ideas, let me give an information-theoretic proof that there are infinitely many prime numbers (Chaitin, 1979). Suppose on the contrary that there are only finitely many primes, in fact, K of them. Consider an algorithmically random natural number N. On the one hand, we know that $I(N)$ is equal to $\log_2 N +$ order of $\log \log N$, since the base-two numeral for N is an algorithmically random $(\log_2 N)$-bit string. On the other hand, N can be calculated from the exponents in its prime factorization, and vice versa. Thus $I(N)$ is equal to the joint information of the K exponents in its prime factorization. By subadditivity, this joint information is bounded from above by the sum of the information contents of the K individual exponents. Each exponent is of order $\log N$. The information content of each exponent is thus of order $\log \log N$. Hence $I(N)$ is simultaneously equal to $\log_2 N + O(\log \log N)$ and less than or equal to $KO(\log \log N)$, which is impossible.

The concepts of algorithmic information theory are made to order for obtaining quantitative incompleteness theorems, and I will now give a number of information-theoretic proofs of Gödel's theorem (Chaitin, 1974a, 1974b, 1975a, 1977, 1982; Chaitin and Schwartz, 1978; Gardner, 1979).

4. Information-Theoretic Proofs of Gödel's Theorem

I propose that we consider a formal axiomatic system to be a computer program for listing the set of theorems, and measure its size in bits. In other words, the measure of the size of a formal axiomatic system that I will use is quite crude. It is merely the amount of space it takes to specify a proof-checking algorithm and how to apply it to all possible proofs, which is roughly the amount of space it takes to be very precise about the alphabet, vocabulary, grammar, axioms, and rules of inference. This is roughly proportional to the number of pages it takes to present the formal axiomatic system in a textbook.

Here is the first information-theoretic incompleteness theorem. Consider an N-bit formal axiomatic system. There is a computer program of size N which does not halt, but one cannot prove this within the formal axiomatic system. On the other hand, N bits of axioms can permit one to deduce precisely which programs of size less than N halt and which ones do not. Here are two different N-bit axioms which do this. If God tells one how many different programs of size less than N halt, this can be expressed as an

N-bit base-two numeral, and from it one could eventually deduce which of these programs halt and which do not. An alternative divine revelation would be knowing that program of size less than N which takes longest to halt. (In the current context, programs have all input contained within them.)

Another way to thwart an N-bit formal axiomatic system is to merely toss an unbiased coin slightly more than N times. It is almost certain that the resulting binary string will be algorithmically random, but it is not possible to prove this within the formal axiomatic system. If one believes the postulate of quantum mechanics that God plays dice with the universe (Albert Einstein did not), then physics provides a means to expose the limitations of formal axiomatic systems. In fact, within an N-bit formal axiomatic system it is not even possible to prove that a particular object has algorithmic information content greater than N, even though almost all (all but finitely many) objects have this property.

The proof of this closely resembles G. G. Berry's paradox of "the first natural number which cannot be named in less than a billion words," published by Russell at the turn of the century (Russell, 1967). The version of Berry's paradox that will do the trick is "that object having the shortest proof that its algorithmic information content is greater than a billion bits." More precisely, "that object having the shortest proof within the following formal axiomatic system that its information content is greater than the information content of the formal axiomatic system: ...," where the dots are to be filled in with a complete description of the formal axiomatic system in question.

By the way, the fact that in a given formal axiomatic system one can only prove that finitely many specific strings are random, is closely related to Post's notion of a simple r.e. set. Indeed, the set of nonrandom or compressible strings is a simple r.e. set. So Berry and Post had the germ of my incompleteness theorem!

In order to proceed, I must define a fascinating algorithmically random real number between zero and one, which I like to call Ω (Chaitin, 1975b; Gardner, 1979). Ω is a suitable subject for worship by mystical cultists, for as Charles Bennett (Gardner, 1979) has argued persuasively, in a sense Ω contains all constructive mathematical truth, and expresses it as concisely and compactly as possible. Knowing the numerical value of Ω with N bits of precision, that is to say, knowing the first N bits of Ω's base-two expansion, is another N-bit axiom that permits one to deduce precisely which programs of size less than N halt and which ones do not.

Ω is defined as the halting probability of whichever standard general-purpose computer has been chosen, if each bit of its program is produced by an independent toss of a fair coin. To Turing's theorem in recursive function theory that the halting problem is unsolvable, there corresponds in algorithmic information theory the theorem that the base-two expansion of Ω is algorithmically random. Therefore it takes N bits of axioms to be able to prove what the first N bits of Ω are, and these bits seem completely accidental like the products of a random physical process. One can therefore measure the power of a formal axiomatic system by how much of the numerical value of Ω it is possible to deduce from its axioms. This is sort of like measuring the power of a formal axiomatic system in terms of the size in bits of the shortest program whose halting problem is undecidable within the formal axiomatic system.

It is possible to dress this incompleteness theorem involving Ω so that no direct mention is made of halting probabilities, in fact, in rather straightforward number-theoretic terms making no mention of computer programs at all. Ω can be represented as the limit of a monotone increasing computable sequence of rational numbers. Its Nth bit is therefore the limit as T tends to infinity of a computable function of N and T. Thus the Nth bit of Ω can be expressed in the form $\exists X \forall Y$[computable predicate of X, Y, and N]. Complete chaos is only two quantifiers away from computability! Ω can also be expressed via a polynomial P in, say, one hundred variables, with integer coefficients and exponents (Davis et al., 1976): the Nth bit of Ω is a 1 if and only if there are infinitely many natural numbers K such that the equation $P(N, K, X_1, \ldots, X_{98}) = 0$ has a solution in natural numbers.

Of course, Ω has the very serious problem that it takes much too long to deduce theorems from it, and this is also the case with the other two axioms we considered. So the ideal, perfect mathematical axiom is in fact useless! One does not really want the most compact axiom for deducing a given set of assertions. Just as there is a trade-off between program size and running time, there is a trade-off between the number of bits of axioms one assumes and the size of proofs. Of course, random or irreducible truths cannot be compressed into axioms shorter than themselves. If, however, a set of assertions is not algorithmically independent, then it takes fewer bits of axioms to deduce them all than the sum of the number of bits of axioms it takes to deduce them separately, and this is desirable as long as the proofs do not get too long. This suggests a pragmatic attitude toward mathematical truth, somewhat more like that of physicists.

Ours has indeed been a long stroll through a gallery of incompleteness theorems. What is the conclusion or moral? It is time to make a final statement about the meaning of Gödel's theorem.

5. The Meaning of Gödel's Theorem

Information theory suggests that the Gödel phenomenon is natural and widespread, not pathological and unusual. Strangely enough, it does this via counting arguments, and without exhibiting individual assertions which are true but unprovable! Of course, it would help to have more proofs that particular interesting and natural true assertions are not demonstrable within fashionable formal axiomatic systems.

The real question is this: Is Gödel's theorem a mandate for revolution, anarchy, and license?! Can one give up after trying for two months to prove a theorem, and add it as a new axiom? This sounds ridiculous, but it is sort of what number theorists have done with Bernhard Riemann's ζ conjecture (Pólya, 1959). Of course, two months is not enough. New axioms should be chosen with care, because of their usefulness and large amounts of evidence suggesting that they are correct, in the same careful manner, say, in practice in the physics community.

Gödel himself has espoused this view with remarkable vigor and clarity, in his discussion of whether Cantor's continuum hypothesis should be added to set theory as a new axiom (Gödel, 1964):

> ...even disregarding the intrinsic necessity of some new axiom, and even in case it has no intrinsic necessity at all, a probable decision about its truth is possible also in another way, namely, inductively by studying its "success." Success here means fruitfulness in consequences, in particular in "verifiable" consequences, i.e., consequences demonstrable without the new axiom, whose proofs with the help of the new axiom, however, are considerably simpler and easier to discover, and make it possible to contract into one proof many different proofs. The axioms for the system of real numbers, rejected by intuitionists, have in this sense been verified to some extent, owing to the fact that analytical number theory frequently allows one to prove number-theoretical theorems which, in a more cumbersome way, can subsequently be

verified by elementary methods. A much higher degree of verification than that, however, is conceivable. There might exist axioms so abundant in their verifiable consequences, shedding so much light upon a whole field, and yielding such powerful methods for solving problems (and even solving them constructively, as far as that is possible) that, no matter whether or not they are intrinsically necessary, they would have to be accepted at least in the same sense as any well-established physical theory.

Later in the same discussion Gödel refers to these ideas again:

> It was pointed out earlier... that, besides mathematical intuition, there exists another (though only probable) criterion of the truth of mathematical axioms, namely their fruitfulness in mathematics and, one may add, possibly also in physics... The simplest case of an application of the criterion under discussion arises when some... axiom has number-theoretical consequences verifiable by computation up to any given integer.

Gödel also expresses himself in no uncertain terms in a discussion of Russell's mathematical logic (Gödel, 1964):

> The analogy between mathematics and a natural science is enlarged upon by Russell also in another respect... axioms need not be evident in themselves, but rather their justification lies (exactly as in physics) in the fact that they make it possible for these "sense perceptions" to be deduced... I think that... this view has been largely justified by subsequent developments, and it is to be expected that it will be still more so in the future. It has turned out that the solution of certain arithmetical problems requires the use of assumptions essentially transcending arithmetic... Furthermore it seems likely that for deciding certain questions of abstract set theory and even for certain related questions of the theory of real numbers new axioms based on some hitherto unknown idea will be necessary. Perhaps also the apparently insurmountable difficulties which some other mathematical problems have been presenting for many years are due to the fact that the necessary axioms have not yet been found. Of course, under these circumstances mathematics may lose a good deal of

its "absolute certainty;" but, under the influence of the modern criticism of the foundations, this has already happened to a large extent...

I end as I began, with a quotation from Weyl (1949): "A truly realistic mathematics should be conceived, in line with physics, as a branch of the theoretical construction of the one real world, and should adopt the same sober and cautious attitude toward hypothetic extensions of its foundations as is exhibited by physics."

6. Directions for Future Research

a. Prove that a famous mathematical conjecture is unsolvable in the usual formalizations of number theory. Problem: if Pierre Fermat's "last theorem" is undecidable then it is true, so this is hard to do.

b. Formalize all of college mathematics in a practical way. One wants to produce textbooks that can be run through a practical formal proof checker and that are not too much larger than the usual ones. LISP (Levin, 1974) and SETL (Dewar et al., 1981) might be good for this.

c. Is algorithmic information theory relevant to physics, in particular, to thermodynamics and statistical mechanics? Explore the thermodynamics of computation (Bennett, 1982) and determine the ultimate physical limitations of computers.

d. Is there a physical phenomenon that computes something noncomputable? Contrariwise, does Turing's thesis that anything computable can be computed by a Turing machine constrain the physical universe we are in?

e. Develop measures of self-organization and formal proofs that life must evolve (Chaitin, 1979; Eigen and Winkler, 1981; von Neumann, 1966).

f. Develop formal definitions of intelligence and measures of its various components; apply information theory and complexity theory to AI.

References

Let me give a few pointers to the literature. The following are my previous publications on Gödel's theorem: Chaitin, 1974a, 1974b, 1975a, 1977, 1982; Chaitin and Schwartz, 1978. Related publications by other authors include Davis, 1978; Gardner, 1979; Hofstadter, 1979; Levin, 1974; Post, 1965. For discussions of the epistemology of mathematics and science, see Einstein, 1944, 1954; Feynman, 1965; Gödel, 1964; Pólya, 1959; von Neumann, 1956, 1963; Taub, 1961; Weyl, 1946, 1949.

- Bell, E. T. (1951). *Mathematics, Queen and Servant of Science*, McGraw-Hill, New York.

- Bennett, C. H. (1982). The thermodynamics of computation—a review, *International Journal of Theoretical Physics*, **21**, 905–940.

- Chaitin, G. J. (1974a). Information-theoretic computational complexity, *IEEE Transactions on Information Theory*, **IT-20**, 10–15.

- Chaitin, G. J. (1974b). Information-theoretic limitations of formal systems, *Journal of the ACM*, **21**, 403–424.

- Chaitin, G. J. (1975a). Randomness and mathematical proof, *Scientific American*, **232** (5) (May 1975), 47–52. (Also published in the French, Japanese, and Italian editions of *Scientific American*.)

- Chaitin, G. J. (1975b). A theory of program size formally identical to information theory, *Journal of the ACM*, **22**, 329–340.

- Chaitin, G. J. (1977). Algorithmic information theory, *IBM Journal of Research and Development*, **21**, 350–359, 496.

- Chaitin, G. J., and Schwartz, J. T. (1978). A note on Monte Carlo primality tests and algorithmic information theory, *Communications on Pure and Applied Mathematics*, **31**, 521–527.

- Chaitin, G. J. (1979). Toward a mathematical definition of "life," in *The Maximum Entropy Formalism*, R. D. Levine and M. Tribus (eds.), MIT Press, Cambridge, Massachusetts, pp. 477–498.

- Chaitin, G. J. (1982). Algorithmic information theory, *Encyclopedia of Statistical Sciences*, Vol. 1, Wiley, New York, pp. 38–41.

- Cole, C. A., Wolfram, S., et al. (1981). *SMP: a symbolic manipulation program*, California Institute of Technology, Pasadena, California.

- Courant, R., and Robbins, H. (1941). *What is Mathematics?*, Oxford University Press, London.

- Davis, M., Matijasevič, Y., and Robinson, J. (1976). Hilbert's tenth problem. Diophantine equations: positive aspects of a negative solution, in *Mathematical Developments Arising from Hilbert Problems, Proceedings of Symposia in Pure Mathematics*, Vol. XXVII, American Mathematical Society, Providence, Rhode Island, pp. 323–378.

- Davis, M. (1978). What is a computation?, in *Mathematics Today: Twelve Informal Essays,* L. A. Steen (ed.), Springer-Verlag, New York, pp. 241–267.

- Dewar, R. B. K., Schonberg, E., and Schwartz, J. T. (1981). *Higher Level Programming: Introduction to the Use of the Set-Theoretic Programming Language SETL,* Courant Institute of Mathematical Sciences, New York University, New York.

- Eigen, M., and Winkler, R. (1981). *Laws of the Game,* Knopf, New York.

- Einstein, A. (1944). Remarks on Bertrand Russell's theory of knowledge, in *The Philosophy of Bertrand Russell,* P. A. Schilpp (ed.), Northwestern University, Evanston, Illinois, pp. 277–291.

- Einstein, A. (1954). *Ideas and Opinions,* Crown, New York, pp. 18–24.

- Feynman, A. (1965). *The Character of Physical Law,* MIT Press, Cambridge, Massachusetts.

- Gardner, M. (1979). The random number Ω bids fair to hold the mysteries of the universe, Mathematical Games Dept., *Scientific American,* **241** (5) (November 1979), 20–34.

- Gödel, K. (1964). Russell's mathematical logic, and What is Cantor's continuum problem?, in *Philosophy of Mathematics,* P. Benacerraf and H. Putnam (eds.), Prentice-Hall, Englewood Cliffs, New Jersey, pp. 211–232, 258–273.

- Hofstadter, D. R. (1979). *Gödel, Escher, Bach: an Eternal Golden Braid,* Basic Books, New York.

- Levin, M. (1974). *Mathematical Logic for Computer Scientists,* MIT Project MAC report MAC TR-131, Cambridge, Massachusetts.

- Pólya, G. (1959). Heuristic reasoning in the theory of numbers, *American Mathematical Monthly,* **66,** 375–384.

- Post, E. (1965). Recursively enumerable sets of positive integers and their decision problems, in *The Undecidable: Basic Papers on Undecidable Propositions, Unsolvable Problems and Computable Functions,* M. Davis (ed.), Raven Press, Hewlett, New York, pp. 305–337.

- Russell, B. (1967). Mathematical logic as based on the theory of types, in *From Frege to Gödel: A Source Book in Mathematical Logic, 1879–1931,* J. van Heijenoort (ed.), Harvard University Press, Cambridge, Massachusetts, pp. 150–182.

- Taub, A. H. (ed.) (1961). *J. von Neumann—Collected Works,* Vol. I, Pergamon Press, New York, pp. 1–9.

- von Neumann, J. (1956). The mathematician, in *The World of Mathematics,* Vol. 4, J. R. Newman (ed.), Simon and Schuster, New York, pp. 2053–2063.

- von Neumann, J. (1963). The role of mathematics in the sciences and in society, and Method in the physical sciences, in *J. von Neumann—Collected Works,* Vol. VI, A. H. Taub (ed.), McMillan, New York, pp. 477–498.

- von Neumann, J. (1966). *Theory of Self-Reproducing Automata,* A. W. Burks (ed.), University of Illinois Press, Urbana, Illinois.

- Weyl, H. (1946). Mathematics and logic, *American Mathematical Monthly,* **53**, 1–13.

- Weyl, H. (1949). *Philosophy of Mathematics and Natural Science,* Princeton University Press, Princeton, New Jersey.

- Wilf, H. S. (1982). The disk with the college education, *American Mathematical Monthly,* **89**, 4–8.

Randomness in arithmetic

It is impossible to prove whether each member of a family of algebraic equations has a finite or an infinite number of solutions: the answers vary randomly and therefore elude mathematical reasoning.

What could be more certain than the fact that 2 plus 2 equals 4? Since the time of the ancient Greeks mathematicians have believed there is little—if anything—as unequivocal as a proved theorem. In fact, mathematical statements that can be proved true have often been regarded as a more solid foundation for a system of thought than any maxim about morals or even physical objects. The 17th-century German mathematician and philosopher Gottfried Wilhelm Leibniz even envisioned a "calculus" of reasoning such that all disputes could one day be settled with the words "Gentlemen, let us compute!" By the beginning of this century symbolic logic had progressed to such an extent that the German mathematician David Hilbert declared that all mathematical questions are in principle decidable, and he confidently set out to codify once and for all the methods of mathematical reasoning.

Such blissful optimism was shattered by the astonishing and profound discoveries of Kurt Gödel and Alan M. Turing in the 1930's. Gödel showed that no finite set of axioms and methods of reasoning could encompass all the mathematical properties of the positive integers. Turing later couched Gödel's ingenious and complicated proof in a more accessible form. He showed that Gödel's incompleteness theorem is equivalent to the assertion that there can be no general method for systematically deciding whether a computer program will ever halt, that is, whether it will ever cause the computer to stop running. Of course, if a particular program does cause the computer to halt, that fact can be easily proved by running the program.

The difficulty lies in proving that an arbitrary program never halts.

I have recently been able to take a further step along the path laid out by Gödel and Turing. By translating a particular computer program into an algebraic equation of a type that was familiar even to the ancient Greeks, I have shown that there is randomness in the branch of pure mathematics known as number theory. My work indicates that—to borrow Einstein's metaphor—God sometimes plays dice with whole numbers!

This result, which is part of a body of work called algorithmic information theory, is not a cause for pessimism; it does not portend anarchy or lawlessness in mathematics. (Indeed, most mathematicians continue working on problems as before.) What it means is that mathematical laws of a different kind might have to apply in certain situations: statistical laws. In the same way that it is impossible to predict the exact moment at which an individual atom undergoes radioactive decay, mathematics is sometimes powerless to answer particular questions. Nevertheless, physicists can still make reliable predictions about averages over large ensembles of atoms. Mathematicians may in some cases be limited to a similar approach.

My work is a natural extension of Turing's, but whereas Turing considered whether or not an arbitrary program would ever halt, I consider the probability that any general-purpose computer will stop running if its program is chosen completely at random. What do I mean when I say "chosen completely at random"? Since at the most fundamental level any program can be reduced to a sequence of bits (each of which can take on the value 0 or 1) that are "read" and "interpreted" by the computer hardware, I mean that a completely random program consisting of n bits could just as well be the result of flipping a coin n times (in which a "heads" represents a 0 and a "tails" represents 1, or vice versa).

The probability that such a completely random program will halt, which I have named omega (Ω), can be expressed in terms of a real number between 0 and 1. (The statement $\Omega = 0$ would mean that no random program will ever halt, and $\Omega = 1$ would mean that every random program halts. For a general-purpose computer neither of these extremes is actually possible.) Because Ω is a real number, it can be fully expressed only as an unending sequence of digits. In base 2 such a sequence would amount to an infinite string of 0's and 1's.

Perhaps the most interesting characteristic of Ω is that it is algorithmically random: it cannot be compressed into a program (considered as a string of bits) shorter than itself. This definition of randomness, which has a central role in algorithmic information theory, was independently formulated in the mid-1960's by the late A. N. Kolmogorov and me. (I have since had to correct the definition.)

The basic idea behind the definition is a simple one. Some sequences of bits can be compressed into programs much shorter than they are, because they follow a pattern or rule. For example, a 200-bit sequence of the form 0101010101... can be greatly compressed by describing it as "100 repetitions of 01." Such sequences certainly are not random. A 200-bit sequence generated by tossing a coin, on the other hand, cannot be compressed, since in general there is no pattern to the succession of 0's and 1's: it is a completely random sequence.

Of all the possible sequences of bits, most are incompressible and therefore random. Since a sequence of bits can be considered to be a base-2 representation of any real number (if one allows infinite sequences), it follows that most real numbers are in fact random. It is not difficult to show that an algorithmically random number, such as Ω, exhibits the usual statistical properties one associates with randomness. One such property is normality: every possible digit appears with equal frequency in the number. In a base-2 representation this means that as the number of digits of Ω approaches infinity, 0 and 1 respectively account for exactly 50 percent of Ω's digits.

A key technical point that must be stipulated in order for Ω to make sense is that an input program must be self-delimiting: its total length (in bits) must be given within the program itself. (This seemingly minor point, which paralyzed progress in the field for nearly a decade, is what entailed the redefinition of algorithmic randomness.) Real programming languages are self-delimiting, because they provide constructs for beginning and ending a program. Such constructs allow a program to contain well-defined subprograms, which may also have other subprograms nested in them. Because a self-delimiting program is built up by concatenating and nesting self-delimiting subprograms, a program is syntactically complete only when the last open subprogram is closed. In essence the beginning and ending constructs for programs and subprograms function respectively like left and right parentheses in mathematical expressions.

If programs were not self-delimiting, they could not be constructed from subprograms, and summing the halting probabilities for all programs would yield an infinite number. If one considers only self-delimiting programs, not only is Ω limited to the range between 0 to 1 but also it can be explicitly calculated "in the limit from below." That is to say, it is possible to calculate an infinite sequence of rational numbers (which can be expressed in terms of a finite sequence of bits) each of which is closer to the true value of Ω than the preceding number.

One way to do this is to systematically calculate Ω_n for increasing values of n; Ω_n is the probability that a completely random program up to n bits in size will halt within n seconds if the program is run on a given computer. Since there are 2^k possible programs that are k bits long, Ω_n can in principle be calculated by determining for every value of k between 1 and n how many of the possible programs actually halt within n seconds, multiplying that number by 2^{-k} and then summing all the products. In other words, each k-bit program that halts contributes 2^{-k} to Ω_n; programs that do not halt contribute 0.

If one were miraculously given the value of Ω with k bits of precision, one could calculate a sequence of Ω_n's until one reached a value that equaled the given value of Ω. At this point one would know all programs of a size less than k bits that halt; in essence one would have solved Turing's halting problem for all programs of a size less than k bits. Of course, the time required for the calculation would be enormous for reasonable values of k.

So far I have been referring exclusively to computers and their programs in discussing the halting problem, but it took on a new dimension in light of the work of J. P. Jones of the University of Calgary and Y. V. Matijasevič of the V. A. Steklov Institute of Mathematics in Leningrad. Their work provides a method for casting the problem as assertions about particular diophantine equations. These algebraic equations, which involve only multiplication, addition and exponentiation of whole numbers, are named after the third-century Greek mathematician Diophantos of Alexandria.

To be more specific, by applying the method of Jones and Matijasevič one can equate the statement that a particular program does not halt with the assertion that one of a particular family of diophantine equations has no solution in whole numbers. As with the original version of the halting problem for computers, it is easy to prove a solution exists: all one has to do is to plug in the correct numbers and verify that the resulting numbers on the left and right sides of the equal sign are in fact equal. The much more difficult problem is to prove that there are absolutely no solutions when this is the case.

The family of equations is constructed from a basic equation that contains a particular variable k, called the parameter, which takes on the values 1, 2, 3 and so on. Hence there is an infinitely large family of equations (one for each value of k) that can be generated from one basic equation for each of a "family" of programs. The mathematical assertion that the diophantine equation with parameter k has no solution encodes the assertion that the kth computer program never halts. On the other hand, if the kth program does halt, then the equation has exactly one solution. In a sense the truth or falsehood of assertions of this type is mathematically uncertain, since it varies unpredictably as the parameter k takes on different values.

My approach to the question of unpredictability in mathematics is similar, but it achieves a much greater degree of randomness. Instead of "arithmetizing" computer programs that may or may not halt as a family of diophantine equations, I apply the method of Jones and Matijasevič to arithmetize a single program to calculate the kth bit in Ω_n.

The method is based on a curious property of the parity of binomial coefficients (whether they are even or odd numbers) that was noticed by Édouard A. Lucas a century ago but was not properly appreciated until now. Binomial coefficients are the multiplicands of the powers of x that arise when one expands expressions of the type $(x + 1)^n$. These coefficients can easily be computed by constructing what is known as Pascal's triangle.

Lucas's theorem asserts that the coefficient of x^k in the expansion of $(x + 1)^n$ is odd only if each digit in the base-2 representation of the number k is less than or equal to the corresponding digit in the base-2 representation of n (starting from the right and reading left). To put it a little more simply, the coefficient for x^k in an expansion of $(x + 1)^n$ is odd if for every bit of k that is a 1 the corresponding bit of n is also a 1, otherwise the coefficient is even. For example, the coefficient of x^2 in the binomial expansion of $(x + 1)^4$ is 6, which is even. Hence the 1 in the base-2 representation of 2 (10) is not matched with a 1 in the same position in the base-2 representation of 4 (100).

Although the arithmetization is conceptually simple and elegant, it is a substantial programming task to carry through the construction. Nevertheless, I thought it would be fun to do it. I therefore developed a "compiler" program for producing equations from programs for a register machine. A register machine is a computer that consists of a small set of registers for storing arbitrarily large numbers. It is an abstraction, of course, since any real computer has registers with a limited capacity.

Feeding a register-machine program that executes instructions in the LISP computer language, as input, into a real computer programmed with the compiler yields within a few minutes, as output, an equation about 200 pages long containing about 17,000 nonnegative integer variables. I can thus derive a diophantine equation having a parameter k that encodes the kth bit of Ω_n merely by plugging a LISP program (in binary form) for calculating the kth bit of Ω_n into the 200-page equation. For any given pair of values of k and n, the diophantine equation has exactly one solution if the kth bit of Ω_n is a 1, and it has no solution if the kth bit of Ω_n is a 0.

Because this applies for any pair of values for k and n, one can in principle keep k fixed and systematically increase the value of n without limit, calculating the kth bit of Ω_n for each value of n. For small values of n the kth bit of Ω_n will fluctuate erratically between 0 and 1. Eventually, however, it will settle on either a 0 or a 1, since for very large values of n it will be equal to the kth bit of Ω, which is immutable. Hence the diophantine equation actually has infinitely many solutions for a particular value of its parameter k if the kth bit of Ω turns out to be a 1, and for similar reasons it has only finitely many solutions if the kth bit of Ω turns out to be a 0. In this way, instead of considering whether a diophantine equation has any solutions for each value of its parameter k, I ask whether it has infinitely many solutions.

Although it might seem that there is little to be gained by asking whether there are infinitely many solutions instead of whether there are any solutions, there is in fact a critical distinction: the answers to my question are logically independent. Two mathematical assertions are logically independent if it is impossible to derive one from the other, that is, if neither is a logical consequence of the other. This notion of independence can usually be distinguished from that applied in statistics. There two chance events are said to be independent if the outcome of one has no bearing on the outcome of the other. For example, the result of tossing a coin in no way affects the result of the next toss: the results are statistically independent.

In my approach I bring both notions of independence to bear. The answer to my question for one value of k is logically independent of the answer for another value of k. The reason is that the individual bits of Ω, which determine the answers, are statistically independent.

Although it is easy to show that for about half of the values of k the number of solutions is finite and for the other half the number of solutions is infinite, there is no possible way to compress the answers in a formula or set of rules; they mimic the results of coin tosses. Because Ω is algorithmically random, even knowing the answers for 1,000 values of k would not help one to give the correct answer for another value of k. A mathematician could do no better than a gambler tossing a coin in deciding whether a particular equation had a finite or an infinite number of solutions. Whatever axioms and proofs one could apply to find the answer for the diophantine equation with one value of k, they would be inapplicable for the same equation with another value of k.

Mathematical reasoning is therefore essentially helpless in such a case, since there are no logical interconnections between the diophantine equations generated in this way. No matter how bright one is or how long the proofs and how complicated the mathematical axioms are, the infinite series of propositions stating whether the number of solutions of the diophantine equations is finite or infinite will quickly defeat one as k increases. Randomness, uncertainty and unpredictability occur even in the elementary branches of number theory that deal with diophantine equations.

How have the incompleteness theorem of Gödel, the halting problem of Turing and my own work affected mathematics? The fact is that most mathematicians have shrugged off the results. Of course, they agree in principle that any finite set of axioms is incomplete, but in practice they dismiss the fact as not applying directly to their work. Unfortunately, however, it may sometimes apply. Although Gödel's original theorem seemed to apply only to unusual mathematical propositions that were not likely to be of interest in practice, algorithmic information theory has shown that incompleteness and randomness are natural and pervasive. This suggests to me that the possibility of searching for new axioms applying to the whole numbers should perhaps be taken more seriously.

Indeed, the fact that many mathematical problems have remained unsolved for hundreds and even thousands of years tends to support my contention. Mathematicians steadfastly assume that the failure to solve these problems lies strictly within themselves, but could the fault not lie in the incompleteness of their axioms? For example, the question of whether there are any perfect odd numbers has defied an answer since the time of the ancient Greeks. (A perfect number is a number that is exactly the sum of its divisors, excluding itself. Hence 6 is a perfect number, since 6 equals 1 plus 2 plus 3.) Could it be that the statement "There are no odd perfect numbers" is unprovable? If it is, perhaps mathematicians had better accept it as an axiom.

This may seem like a ridiculous suggestion to most mathematicians, but to a physicist or a biologist it may not seem so absurd. To those who work in the empirical sciences the usefulness of a hypothesis, and not necessarily its "self-evident truth," is the key criterion by which to judge whether it should be regarded as the basis for a theory. If there are many conjectures that can be settled by invoking a hypothesis, empirical scientists take the hypothesis seriously. (The nonexistence of odd perfect numbers does not appear to have significant implications and would therefore not be a useful axiom by this criterion.)

Actually in a few cases mathematicians have already taken unproved but useful conjectures as a basis for their work. The so-called Riemann hypothesis, for instance, is often accepted as being true, even though it has never been proved, because many other important theorems are based on it. Moreover, the hypothesis has been tested empirically by means of the most powerful computers, and none has come up with a single counterexample. Indeed, computer programs (which, as I have indicated, are equivalent to mathematical statements) are also tested in this way—by verifying a number of test cases rather than by rigorous mathematical proof.

Are there other problems in other fields of science that can benefit from these insights into the foundations of mathematics? I believe algorithmic information theory may have relevance to biology. The regulatory genes of a developing embryo are in effect a computer program for constructing an organism. The "complexity" of this biochemical computer program could conceivably be measured in terms analogous to those I have developed in quantifying the information content of Ω.

Although Ω is completely random (or infinitely complex) and cannot ever be computed exactly, it can be approximated with arbitrary precision given an infinite amount of time. The complexity of living organisms, it seems to me, could be approximated in a similar way. A sequence of Ω_n's, which approach Ω, can be regarded as a metaphor for evolution and perhaps could contain the germ of a mathematical model for the evolution of biological complexity.

At the end of his life John von Neumann challenged mathematicians to find an abstract mathematical theory for the origin and evolution of life. This fundamental problem, like most fundamental problems, is magnificently difficult. Perhaps algorithmic information theory can help to suggest a way to proceed.

GREGORY J. CHAITIN *is on the staff of the IBM Thomas J. Watson Research Center in Yorktown Heights, N.Y. He is the principal architect of algorithmic information theory and has just published two books in which the theory's concepts are applied to elucidate the nature of randomness and the limitations of mathematics. This is Chaitin's second article for* SCIENTIFIC AMERICAN.

Further Reading

- ALGORITHMIC INFORMATION THEORY. Gregory J. Chaitin. Cambridge University Press, 1987.

- INFORMATION, RANDOMNESS & INCOMPLETENESS. Gregory J. Chaitin. World Scientific Publishing Co. Pte. Ltd., 1987.

- THE ULTIMATE IN UNDECIDABILITY. Ian Stewart in *Nature,* Vol. 232, No. 6160, pages 115–116; March 10, 1988.

Randomness in arithmetic and the decline & fall of reductionism in pure mathematics

Lecture given Thursday 22 October 1992 at a Mathematics – Computer Science Colloquium at the University of New Mexico. The lecture was videotaped; this is an edited transcript.

1. Hilbert on the axiomatic method

Last month I was a speaker at a symposium on reductionism at Cambridge University where Turing did his work. I'd like to repeat the talk I gave there and explain how my work continues and extends Turing's. Two previous speakers had said bad things about David Hilbert. So I started by saying that in spite of what you might have heard in some of the previous lectures, Hilbert was not a twit!

Hilbert's idea is the culmination of two thousand years of mathematical tradition going back to Euclid's axiomatic treatment of geometry, going back to Leibniz's dream of a symbolic logic and Russell and Whitehead's monumental *Principia Mathematica*. Hilbert's dream was to once and for all clarify the methods of mathematical reasoning. Hilbert wanted to formulate a formal axiomatic system which would encompass all of mathematics.

Formal Axiomatic System

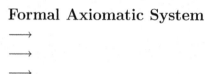

Hilbert emphasized a number of key properties that such a formal axiomatic system should have. It's like a computer programming language. It's a precise statement about the methods of reasoning, the postulates and the methods of inference that we accept as mathematicians. Furthermore Hilbert stipulated that the formal axiomatic system encompassing all of mathematics that he wanted to construct should be "consistent" and it should be "complete."

Formal Axiomatic System

$\longrightarrow$ consistent
$\longrightarrow$ complete
$\longrightarrow$

Consistent means that you shouldn't be able to prove an assertion and the contrary of the assertion.

Formal Axiomatic System

$\longrightarrow$ consistent $A \neg A$
$\longrightarrow$ complete
$\longrightarrow$

You shouldn't be able to prove A and not A. That would be very embarrassing.

Complete means that if you make a meaningful assertion you should be able to settle it one way or the other. It means that either A or not A should be a theorem, should be provable from the axioms using the rules of inference in the formal axiomatic system.

Formal Axiomatic System

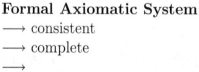

Consider a meaningful assertion A and its contrary not A. Exactly one of the two should be provable if the formal axiomatic system is consistent and complete.

A formal axiomatic system is like a programming language. There's an alphabet and rules of grammar, in other words, a formal syntax. It's a kind of thing that we are familiar with now. Look back at Russell and Whitehead's three enormous volumes full of symbols and you'll feel you're looking at a large computer program in some incomprehensible programming language.

Now there's a very surprising fact. Consistent and complete means only truth and all the truth. They seem like reasonable requirements. There's a funny consequence, though, having to do with something called the decision problem. In German it's the Entscheidungsproblem.

Formal Axiomatic System
$\longrightarrow$ consistent $A \neg A$
$\longrightarrow$ complete $A \neg A$
$\longrightarrow$ decision problem

Hilbert ascribed a great deal of importance to the decision problem.

HILBERT
Formal Axiomatic System
$\longrightarrow$ consistent $A \neg A$
$\longrightarrow$ complete $A \neg A$
$\longrightarrow$ decision problem

Solving the decision problem for a formal axiomatic system is giving an algorithm that enables you to decide whether any given meaningful assertion is a theorem or not. A solution of the decision problem is called a decision procedure.

HILBERT
Formal Axiomatic System
$\longrightarrow$ consistent $A \neg A$
$\longrightarrow$ complete $A \neg A$
$\longrightarrow$ decision procedure

This sounds weird. The formal axiomatic system that Hilbert wanted to construct would have included all of mathematics: elementary arithmetic, calculus, algebra, everything. If there's a decision procedure, then mathematicians are out of work. This algorithm, this mechanical procedure, can check whether something is a theorem or not, can check whether it's true or not. So to require that there be a decision procedure for this formal axiomatic system sounds like you're asking for a lot.

However it's very easy to see that if it's consistent and it's complete that implies that there must be a decision procedure. Here's how you do it. You have a formal language with a finite alphabet and a grammar. And Hilbert emphasized that the whole point of a formal axiomatic system is that there must be a mechanical procedure for checking whether a purported proof is correct or not, whether it obeys the rules or not. That's the notion that mathematical truth should be objective so that everyone can agree whether a proof follows the rules or not.

So if that's the case you run through all possible proofs in size order, and look at all sequences of symbols from the alphabet one character long, two, three, four, a thousand, a thousand and one... a hundred thousand characters long. You apply the mechanical procedure which is the essence of the formal axiomatic system, to check whether each proof is valid. Most of the time, of course, it'll be nonsense, it'll be ungrammatical. But you'll eventually find every possible proof. It's like a million monkeys typing away. You'll find every possible proof, though only in principle of course. The number grows exponentially and this is something that you couldn't do in practice. You'd never get to proofs that are one page long.

But in principle you could run through all possible proofs, check which ones are valid, see what they prove, and that way you can systematically find all theorems. In other words, there is an algorithm, a mechanical procedure, for generating one by one every theorem that can be demonstrated in a formal axiomatic system. So if for every meaningful assertion within the system, either the assertion is a theorem or its contrary is a theorem, only one of them, then you get a decision procedure. To see whether an assertion is a theorem or not you just run through all possible proofs until you find the assertion coming out as a theorem or you prove the contrary assertion.

So it seems that Hilbert actually believed that he was going to solve once and for all, all mathematical problems. It sounds amazing, but apparently he did. He believed that he would be able to set down a consistent and complete formal axiomatic system for all of mathematics and from it obtain a decision procedure for all of mathematics. This is just following the formal, axiomatic tradition in mathematics.

But I'm sure he didn't think that it would be a practical decision procedure. The one I've outlined would only work in principle. It's exponentially slow, it's terribly slow! Totally impractical. But the idea was that if all mathematicians could agree whether a proof is correct and be consistent and complete, in principle that would give a decision procedure for automatically solving any mathematical problem. This was Hilbert's magnificent dream, and it was to be the culmination of Euclid and Leibniz, and Boole and Peano, and Russell and Whitehead.

Of course the only problem with this inspiring project is that it turned out to be impossible!

2. Gödel, Turing and Cantor's diagonal argument

Hilbert is indeed inspiring. His famous lecture in the year 1900 is a call to arms to mathematicians to solve a list of twenty-three difficult problems. As a young kid becoming a mathematician you read that list of twenty-three problems and Hilbert is saying that there is no limit to what mathematicians can do. We can solve a problem if we are clever enough and work at it long enough. He didn't believe that in principle there was any limit to what mathematics could achieve.

I think this is very inspiring. So did John von Neumann. When he was a young man he tried to carry through Hilbert's ambitious program. Because Hilbert couldn't quite get it all to work, in fact he started off just with elementary number theory, 1, 2, 3, 4, 5, ..., not even with real numbers at first.

And then in 1931 to everyone's great surprise (including von Neumann's), Gödel showed that it was impossible, that it couldn't be done, as I'm sure you all know.

Gödel 1931

This was the opposite of what everyone had expected. Von Neumann said it never occurred to him that Hilbert's program couldn't be carried out. Von Neumann admired Gödel enormously, and helped him to get a permanent position at the Institute for Advanced Study.

What Gödel showed was the following. Suppose that you have a formal axiomatic system dealing with elementary number theory, with 1, 2, 3, 4, 5 and addition and multiplication. And we'll assume that it's consistent, which is a minimum requirement—if you can prove false results it's really pretty bad. What Gödel showed was that if you assume that it's consistent, then you can show that it's incomplete. That was Gödel's result, and the proof is very clever and involves self-reference. Gödel was able to construct an assertion about the whole numbers that says of itself that it's unprovable. This was a tremendous shock. Gödel has to be admired for his intellectual imagination; everyone else thought that Hilbert was right.

However I think that Turing's 1936 approach is better.

Gödel 1931
Turing 1936

Gödel's 1931 proof is very ingenious, it's a real tour de force. I have to confess that when I was a kid trying to understand it, I could read it and follow it step by step but somehow I couldn't ever really feel that I was grasping it. Now Turing had a completely different approach.

Turing's approach I think it's fair to say is in some ways more fundamental. In fact, Turing did more than Gödel. Turing not only got as a corollary Gödel's result, he showed that there could be no decision procedure.

You see, if you assume that you have a formal axiomatic system for arithmetic and it's consistent, from Gödel you know that it can't be complete, but there still might be a decision procedure. There still might be a mechanical procedure which would enable you to decide if a given assertion is true or not. That was left open by Gödel, but Turing settled it. The fact that there cannot be a decision procedure is more fundamental and you get incompleteness as a corollary.

How did Turing do it? I want to tell you how he did it because that's the springboard for my own work. The way he did it, and I'm sure all of you have heard about it, has to do with something called the halting problem. In fact if you go back to Turing's 1936 paper you will not find the words "halting problem." But the idea is certainly there.

People also forget that Turing was talking about "computable numbers." The title of his paper is "On computable numbers, with an application to the Entscheidungsproblem." Everyone remembers that the halting problem is unsolvable and that comes from that paper, but not as many people remember that Turing was talking about computable real numbers. My work deals with computable and dramatically uncomputable real numbers. So I'd like to refresh your memory how Turing's argument goes.

Turing's argument is really what destroys Hilbert's dream, and it's a simple argument. It's just Cantor's diagonal procedure (for those of you who know what that is) applied to the computable real numbers. That's it, that's the whole idea in a nutshell, and it's enough to show that Hilbert's dream, the culmination of two thousand years of what mathematicians thought mathematics was about, is wrong. So Turing's work is tremendously deep.

What is Turing's argument? A real number, you know $3.1415926\cdots$, is a length measured with arbitrary precision, with an infinite number of digits. And a computable real number said Turing is one for which there is a computer program or algorithm for calculating the digits one by one. For example, there are programs for π, and there are algorithms for solutions of algebraic equations with integer coefficients. In fact most of the numbers that you actually find in analysis are computable. However they're the exception, if you know set theory, because the computable reals are denumerable and the reals are nondenumerable (you don't have to know what that means). That's the essence of Turing's idea.

The idea is this. You list all possible computer programs. At that time there were no computer programs, and Turing had to invent the Turing machine, which was a tremendous step forward. But now you just say, imagine writing a list with every possible computer program.

$$
\begin{array}{ll}
p_1 & \textbf{Gödel } 1931 \\
p_2 & \textbf{Turing } 1936 \\
p_3 & \\
p_4 & \\
p_5 & \\
p_6 & \\
\vdots &
\end{array}
$$

If you consider computer programs to be in binary, then it's natural to think of a computer program as a natural number. And next to each computer program, the first one, the second one, the third one, write out the real number that it computes if it computes a real (it may not). But if it prints out an infinite number of digits, write them out. So maybe it's 3.1415926 and here you have another and another and another:

$$
\begin{array}{lll}
p_1 & 3.1415926\cdots & \textbf{Gödel } 1931 \\
p_2 & \cdots & \textbf{Turing } 1936 \\
p_3 & \cdots & \\
p_4 & \cdots & \\
p_5 & \cdots & \\
p_6 & \cdots & \\
\vdots & &
\end{array}
$$

So you make this list. Maybe some of these programs don't print out an infinite number of digits, because they're programs that halt or that have an error in them and explode. But then there'll just be a blank line in the list.

$$p_1 \quad 3.1415926\cdots \qquad \textbf{Gödel } 1931$$
$$p_2 \quad \cdots \qquad\qquad\quad \textbf{Turing } 1936$$
$$p_3 \quad \cdots$$
$$p_4 \quad \cdots$$
$$p_5$$
$$p_6 \quad \cdots$$
$$\vdots$$

It's not really important—let's forget about this possibility.

Following Cantor, Turing says go down the diagonal and look at the first digit of the first number, the second digit of the second, the third...

$$p_1 \quad -.\underline{d_{11}}d_{12}d_{13}d_{14}d_{15}d_{16}\cdots \qquad \textbf{Gödel } 1931$$
$$p_2 \quad -.d_{21}\underline{d_{22}}d_{23}d_{24}d_{25}d_{26}\cdots \qquad \textbf{Turing } 1936$$
$$p_3 \quad -.d_{31}d_{32}\underline{d_{33}}d_{34}d_{35}d_{36}\cdots$$
$$p_4 \quad -.d_{41}d_{42}d_{43}\underline{d_{44}}d_{45}d_{46}\cdots$$
$$p_5$$
$$p_6 \quad -.d_{61}d_{62}d_{63}d_{64}d_{65}\underline{d_{66}}\cdots$$
$$\vdots$$

Well actually it's the digits after the decimal point. So it's the first digit after the decimal point of the the first number, the second digit after the decimal point of the second, the third digit of the third number, the fourth digit of the fourth, the fifth digit of the fifth. And it doesn't matter if the fifth program doesn't put out a fifth digit, it really doesn't matter.

What you do is you change these digits. Make them different. Change every digit on the diagonal. Put these changed digits together into a new number with a decimal point in front, a new real number. That's Cantor's diagonal procedure. So you have a digit which you choose to be different from the first digit of the first number, the second digit of the second, the third of the third, and you put these together into one number.

$$p_1 \quad -.\underline{d_{11}}d_{12}d_{13}d_{14}d_{15}d_{16}\cdots$$
$$p_2 \quad -.d_{21}\underline{d_{22}}d_{23}d_{24}d_{25}d_{26}\cdots$$
$$p_3 \quad -.d_{31}d_{32}\underline{d_{33}}d_{34}d_{35}d_{36}\cdots$$
$$p_4 \quad -.d_{41}d_{42}d_{43}\underline{d_{44}}d_{45}d_{46}\cdots$$
$$p_5$$
$$p_6 \quad -.d_{61}d_{62}d_{63}d_{64}d_{65}\underline{d_{66}}\cdots$$
$$\vdots$$

Gödel 1931

Turing 1936

$$. \neq d_{11} \neq d_{22} \neq d_{33} \neq d_{44} \neq d_{55} \neq d_{66}\cdots$$

This new number cannot be in the list because of the way it was constructed. Therefore it's an uncomputable real number. How does Turing go on from here to the halting problem? Well, just ask yourself **why** can't you compute it? I've explained how to get this number and it looks like you could almost do it. To compute the Nth digit of this number, you get the Nth computer program (you can certainly do that) and then you start it running until it puts out an Nth digit, and at that point you change it. Well what's the problem? That sounds easy.

The problem is, what happens if the Nth computer program never puts out an Nth digit, and you sit there waiting? And that's the halting problem—you cannot decide whether the Nth computer program will ever put out an Nth digit! This is how Turing got the unsolvability of the halting problem. Because if you could solve the halting problem, then you could decide if the Nth computer program ever puts out an Nth digit. And if you could do that then you could actually carry out Cantor's diagonal procedure and compute a real number which has to differ from any computable real. That's Turing's original argument.

Why does this explode Hilbert's dream? What has Turing proved? That there is no algorithm, no mechanical procedure, which will decide if the Nth computer program ever outputs an Nth digit. Thus there can be no algorithm which will decide if a computer program ever halts (finding the Nth digit put out by the Nth program is a special case). Well, what Hilbert wanted was a formal axiomatic system from which all mathematical truth should follow, only mathematical truth, and all mathematical truth. If Hilbert could do that, it would give us a mechanical procedure to decide if a computer program will ever halt. Why?

You just run through all possible proofs until you either find a proof that the program halts or you find a proof that it never halts. So if Hilbert's dream of a finite set of axioms from which all of mathematical truth should follow were possible, then by running through all possible proofs checking which ones are correct, you would be able to decide if any computer program halts. In principle you could. But you **can't** by Turing's very simple argument which is just Cantor's diagonal argument applied to the computable reals. That's how simple it is!

Gödel's proof is ingenious and difficult. Turing's argument is so fundamental, so deep, that everything seems natural and inevitable. But of course he's building on Gödel's work.

3.   The halting probability and algorithmic randomness

The reason I talked to you about Turing and computable reals is that I'm going to use a different procedure to construct an uncomputable real, a much more uncomputable real than Turing does.

$$p_1 \quad -.\underline{d_{11}}d_{12}d_{13}d_{14}d_{15}d_{16}\cdots$$

$$p_2 \quad -.d_{21}\underline{d_{22}}d_{23}d_{24}d_{25}d_{26}\cdots$$

$$p_3 \quad -.d_{31}d_{32}\underline{d_{33}}d_{34}d_{35}d_{36}\cdots$$

$$p_4 \quad -.d_{41}d_{42}d_{43}\underline{d_{44}}d_{45}d_{46}\cdots$$

$$p_5$$

$$p_6 \quad -.d_{61}d_{62}d_{63}d_{64}d_{65}\underline{d_{66}}\cdots$$

$$\vdots$$

Gödel 1931
Turing 1936
uncomputable reals

$$.\neq d_{11} \neq d_{22} \neq d_{33} \neq d_{44} \neq d_{55} \neq d_{66}\cdots$$

And that's how we're going to get into much worse trouble.

How do I get a much more uncomputable real? (And I'll have to tell you how uncomputable it is.) Well, not with Cantor's diagonal argument. I get this number, which I like to call Ω, like this:

$$\Omega = \sum_{p \text{ halts}} 2^{-|p|}$$

This is just the halting probability. It's sort of a mathematical pun. Turing's fundamental result is that the halting problem is unsolvable—there is no algorithm that'll settle the halting problem. My fundamental result is that the halting probability is algorithmically irreducible or algorithmically random.

What exactly is the halting probability? I've written down an expression for it:

$$\Omega = \sum_{p \text{ halts}} 2^{-|p|}$$

Instead of looking at individual programs and asking whether they halt, you put all computer programs together in a bag. If you generate a computer program at random by tossing a coin for each bit of the program, what is the chance that the program will halt? You're thinking of programs as bit strings, and you generate each bit by an independent toss of a fair coin, so if a program is N bits long, then the probability that you get that particular program is 2^{-N}. Any program p that halts contributes $2^{-|p|}$, two to the minus its size in bits, the number of bits in it, to this halting probability.

By the way there's a technical detail which is very important and didn't work in the early version of algorithmic information theory. You couldn't write this:

$$\Omega = \sum_{p \text{ halts}} 2^{-|p|}$$

It would give infinity. The technical detail is that no extension of a valid program is a valid program. Then this sum

$$\sum_{p \text{ halts}} 2^{-|p|}$$

turns out to be between zero and one. Otherwise it turns out to be infinity. It only took ten years until I got it right. The original 1960s version of algorithmic information theory is wrong. One of the reasons it's wrong is that you can't even define this number

$$\Omega = \sum_{p \text{ halts}} 2^{-|p|}$$

In 1974 I redid algorithmic information theory with "self-delimiting" programs and then I discovered the halting probability Ω.

Okay, so this is a probability between zero and one

$$0 < \Omega = \sum_{p \text{ halts}} 2^{-|p|} < 1$$

like all probabilities. The idea is you generate each bit of a program by tossing a coin and ask what is the probability that it halts. This number Ω, this halting probability, is not only an uncomputable real—Turing already knew how to do that. It is uncomputable in the worst possible way. Let me give you some clues how uncomputable it is.

Well, one thing is it's algorithmically incompressible. If you want to get the first N bits of Ω out of a computer program, if you want a computer program that will print out the first N bits of Ω and then halt, that computer program has to be N bits long. Essentially you're only printing out constants that are in the program. You cannot squeeze the first N bits of Ω. This

$$0 < \Omega = \sum_{p \text{ halts}} 2^{-|p|} < 1$$

is a real number, you could write it in binary. And if you want to get out the first N bits from a computer program, essentially you just have to put them in. The program has to be N bits long. That's irreducible algorithmic information. There is no concise description.

Now that's an abstract way of saying things. Let me give a more concrete example of how random Ω is. Émile Borel at the turn of this century was one of the founders of probability theory, and he talked about something he called a normal number.

$$0 < \Omega = \sum_{p \text{ halts}} 2^{-|p|} < 1$$

Émile Borel — normal reals

What is a normal real number? People have calculated π out to a billion digits, maybe two billion. One of the reasons for doing this, besides that it's like climbing a mountain and having the world record, is the question of whether each digit occurs the same number of times. It looks like the digits 0 through 9 each occur 10% of the time in the decimal expansion of π. It looks that way, but nobody can prove it. I think the same is true for $\sqrt{2}$, although that's not as popular a number to ask this about.

Let me describe some work Borel did around the turn of the century when he was pioneering modern probability theory. Pick a real number in the unit interval, a real number with a decimal point in front, with no integer part. If you pick a real number in the unit interval, Borel showed that with probability one it's going to be "normal." Normal means that when you write it in decimal each digit will occur in the limit exactly 10% of the time, and this will also happen in any other base. For example in binary 0 and 1 will each occur in the limit exactly 50% of the time. Similarly with blocks of digits. This was called an absolutely normal real number by Borel, and he showed that with probability one if you pick a real number at random between zero and one it's going to have this property. There's only one problem. He didn't know whether π is normal, he didn't know whether $\sqrt{2}$ is normal. In fact, he couldn't exhibit a single individual example of a normal real number.

The first example of a normal real number was discovered by a friend of Alan Turing's at Cambridge called David Champernowne, who is still alive and who's a well-known economist. Turing was impressed with him—I think he called him "Champ"—because Champ had published this in a paper as an undergraduate. This number is known as Champernowne's number. Let me show you Champernowne's number.

$$0 < \Omega = \sum_{p \text{ halts}} 2^{-|p|} < 1$$

Émile Borel — normal reals
Champernowne
$.01234567891011121314 \cdots 99100101 \cdots$

It goes like this. You write down a decimal point, then you write 0, 1, 2, 3, 4, 5, 6, 7, 8, 9, then 10, 11, 12, 13, 14 until 99, then 100, 101. And you keep going in this funny way. This is called Champernowne's number and Champernowne showed that it's normal in base ten, only in base ten. Nobody knows if it's normal in other bases, I think it's still open. In base ten though, not only will the digits 0 through 9 occur exactly 10% of the time in the limit, but each possible block of two digits will occur exactly 1% of the time in the limit, each block of three digits will occur exactly .1% of the time in the limit, etc. That's called being normal in base ten. But nobody knows what happens in other bases.

The reason I'm saying all this is because it follows from the fact that the halting probability Ω is algorithmically irreducible information that this

$$0 < \Omega = \sum_{p \text{ halts}} 2^{-|p|} < 1$$

is normal in any base. That's easy to prove using ideas about coding and compressing information that go back to Shannon. So here we finally have an example of an absolutely normal number. I don't know how natural you think it is, but it is a specific real number that comes up and is normal in the most demanding sense that Borel could think of. Champernowne's number couldn't quite do that.

This number Ω is in fact random in many more senses. I would say it this way. It cannot be distinguished from the result of independent tosses of a fair coin. In fact this number

$$0 < \Omega = \sum_{p \text{ halts}} 2^{-|p|} < 1$$

shows that you have total randomness and chaos and unpredictability and lack of structure in pure mathematics! The same way that all it took for Turing to destroy Hilbert's dream was the diagonal argument, you just write down this expression

$$0 < \Omega = \sum_{p \text{ halts}} 2^{-|p|} < 1$$

and this shows that there are regions of pure mathematics where reasoning is totally useless, where you're up against an impenetrable wall. This is all it takes. It's just this halting probability.

Why do I say this? Well, let's say you want to use axioms to prove what the bits of this number Ω are. I've already told you that it's uncomputable—right?—like the number that Turing constructs using Cantor's diagonal argument. So we know there is no algorithm which will compute digit by digit or bit by bit this number Ω. But let's try to prove what individual bits are using a formal axiomatic system. What happens?

The situation is very, very bad. It's like this. Suppose you have a formal axiomatic system which is N bits of formal axiomatic system (I'll explain what this means more precisely later). It turns out that with a formal axiomatic system of complexity N, that is, N bits in size, you can prove what the positions and values are of at most $N + c$ bits of Ω.

Now what do I mean by formal axiomatic system N bits in size? Well, remember that the essence of a formal axiomatic system is a mechanical procedure for checking whether a formal proof follows the rules or not. It's a computer program. Of course in Hilbert's days there were no computer programs, but after Turing invented Turing machines you could finally specify the notion of computer program exactly, and of course now we're very familiar with it.

So the proof-checking algorithm which is the essence of any formal axiomatic system in Hilbert's sense is a computer program, and just see how many bits long this computer program is.[1] That's essentially how many bits it takes to specify the rules of the game, the axioms and postulates and the rules of inference. If that's N bits, then you may be able to prove say that the first bit of Ω in binary is 0, that the second bit is 1, that the third bit is 0, and then there might be a gap, and you might be able to prove that the thousandth bit is 1. But you're only going to be able to settle N cases if your formal axiomatic system is an N-bit formal axiomatic system.

Let me try to explain better what this means. It means that you can only get out as much as you put in. If you want to prove whether an individual bit in a specific place in the binary expansion of the real number Ω is a 0 or a 1, essentially the only way to prove that is to take it as a hypothesis, as an axiom, as a postulate. It's irreducible mathematical information. That's the key phrase that really gives the whole idea.

Irreducible Mathematical Information

$$0 < \Omega = \sum_{p \text{ halts}} 2^{-|p|} < 1$$

Émile Borel — normal reals
Champernowne
$.01234567891011121314\cdots99100101\cdots$

Okay, so what have we got? We have a rather simple mathematical object that completely escapes us. Ω's bits have no structure. There is no pattern, there is no structure that we as mathematicians can comprehend. If you're interested in proving what individual bits of this number at specific places are, whether they're 0 or 1, reasoning is completely useless. Here mathematical reasoning is irrelevant and can get nowhere. As I said before, the only way a formal axiomatic system can get out these results is essentially just to put them in as assumptions, which means you're not using reasoning. After all, anything can be demonstrated by taking it as a postulate that you add to your set of axioms. So this is a worst possible case—this is irreducible mathematical information. Here is a case where there is no structure, there are no correlations, there is no pattern that we can perceive.

---

[1] *Technical Note:* Actually, it's best to think of the complexity of a formal axiomatic system as the size in bits of the computer program that enumerates the set of all theorems.

4. Randomness in arithmetic

Okay, what does this have to do with randomness in arithmetic? Now we're going back to Gödel—I skipped over him rather quickly, and now let's go back.

Turing says that you cannot use proofs to decide whether a program will halt. You can't always prove that a program will halt or not. That's how he destroys Hilbert's dream of a universal mathematics. I get us into more trouble by looking at a different kind of question, namely, can you prove that the fifth bit of this particular real number

$$0 < \Omega = \sum_{p \text{ halts}} 2^{-|p|} < 1$$

is a 0 or a 1, or that the eighth bit is a 0 or a 1. But these are strange-looking questions. Who had ever heard of the halting problem in 1936? These are not the kind of things that mathematicians normally worry about. We're getting into trouble, but with questions rather far removed from normal mathematics.

Even though you can't have a formal axiomatic system which can always prove whether a program halts or not, it might be good for everything else and then you could have an **amended** version of Hilbert's dream. And the same with the halting probability Ω. If the halting problem looks a little bizarre, and it certainly did in 1936, well, Ω is brand new and certainly looks bizarre. Who ever heard of a halting probability? It's not the kind of thing that mathematicians normally do. So what do I care about all these incompleteness results!

Well, Gödel had already faced this problem with his assertion which is true but unprovable. It's an assertion which says of itself that it's unprovable. That kind of thing also never comes up in real mathematics. One of the key elements in Gödel's proof is that he managed to construct an **arithmetical** assertion which says of itself that it's unprovable. It was getting this self-referential assertion to be in elementary number theory which took so much cleverness.

There's been a lot of work building on Gödel's work, showing that problems involving computations are equivalent to arithmetical problems involving whole numbers. A number of names come to mind. Julia Robinson, Hilary Putnam and Martin Davis did some of the important work, and then a key result was found in 1970 by Yuri Matijasevič. He constructed a dio-phantine equation, which is an algebraic equation involving only whole num-bers, with a lot of variables. One of the variables, K, is distinguished as a parameter. It's a polynomial equation with integer coefficients and all of the unknowns have to be whole numbers—that's a diophantine equation. As I said, one of the unknowns is a parameter. Matijasevič's equation has a solution for a particular value of the parameter K if and only if the Kth computer program halts.

In the year 1900 Hilbert had asked for an algorithm which will decide whether a diophantine equation, an algebraic equation involving only whole numbers, has a solution. This was Hilbert's tenth problem. It was tenth in his famous list of twenty-three problems. What Matijasevič showed in 1970 was that this is equivalent to deciding whether an arbitrary computer program halts. So Turing's halting problem is exactly as hard as Hilbert's tenth problem. It's exactly as hard to decide whether an arbitrary program will halt as to decide whether an arbitrary algebraic equation in whole numbers has a solution. Therefore there is no algorithm for doing that and Hilbert's tenth problem cannot be solved—that was Matijasevič's 1970 result.

Matijasevič has gone on working in this area. In particular there is a piece of work he did in collaboration with James Jones in 1984. I can use it to follow in Gödel's footsteps, to follow Gödel's example. You see, I've shown that there's complete randomness, no pattern, lack of structure, and that reasoning is completely useless, if you're interested in the individual bits of this number

$$0 < \Omega = \sum_{p \text{ halts}} 2^{-|p|} < 1$$

Following Gödel, let's convert this into something in elementary number theory. Because if you can get into all this trouble in elementary number theory, that's the bedrock. Elementary number theory, 1, 2, 3, 4, 5, addition and multiplication, that goes back to the ancient Greeks and it's the most solid part of all of mathematics. In set theory you're dealing with strange objects like large cardinals, but here you're not even dealing with derivatives or integrals or measure, only with whole numbers. And using the 1984 results of Jones and Matijasevič I can indeed dress up Ω arithmetically and get randomness in elementary number theory.

What I get is an exponential diophantine equation with a parameter. "Exponential diophantine equation" just means that you allow variables in the exponents. In contrast, what Matijasevič used to show that Hilbert's tenth problem is unsolvable is just a polynomial diophantine equation, which means that the exponents are always natural number constants. I have to allow X^Y. It's not known yet whether I actually need to do this. It might be the case that I can manage with a polynomial diophantine equation. It's an open question, I believe that it's not settled yet. But for now, what I have is an exponential diophantine equation with seventeen thousand variables. This equation is two-hundred pages long and again one variable is the parameter.

This is an equation where every constant is a whole number, a natural number, and all the variables are also natural numbers, that is, positive integers. (Actually **non-negative** integers.) One of the variables is a parameter, and you change the value of this parameter—take it to be 1, 2, 3, 4, 5. Then you ask, does the equation have a finite or infinite number of solutions? My equation is constructed so that it has a finite number of solutions if a particular individual bit of Ω is a 0, and it has an infinite number of solutions if that bit is a 1. So deciding whether my exponential diophantine equation in each individual case has a finite or infinite number of solutions is exactly the same as determining what an individual bit of this

$$0 < \Omega = \sum_{p \text{ halts}} 2^{-|p|} < 1$$

halting probability is. And this is completely intractable because Ω is irreducible mathematical information.

Let me emphasize the difference between this and Matijasevič's work on Hilbert's tenth problem. Matijasevič showed that there is a polynomial diophantine equation with a parameter with the following property: You vary the parameter and ask, does the equation have a solution? That turns out to be equivalent to Turing's halting problem, and therefore escapes the power of mathematical reasoning, of formal axiomatic reasoning.

How does this differ from what I do? I use an exponential diophantine equation, which means I allow variables in the exponent. Matijasevič only allows constant exponents. The big difference is that Hilbert asked for an algorithm to decide if a diophantine equation has a solution. The question I have to ask to get randomness in elementary number theory, in the arithmetic of the natural numbers, is slightly more sophisticated. Instead of asking whether there is a solution, I ask whether there are a finite or infinite number of solutions—a more abstract question. This difference is necessary.

My two-hundred page equation is constructed so that it has a finite or infinite number of solutions depending on whether a particular bit of the halting probability is a 0 or a 1. As you vary the parameter, you get each individual bit of Ω. Matijasevič's equation is constructed so that it has a solution if and only if a particular program ever halts. As you vary the parameter, you get each individual computer program.

Thus even in arithmetic you can find Ω's absolute lack of structure, Ω's randomness and irreducible mathematical information. Reasoning is completely powerless in those areas of arithmetic. My equation shows that this is so. As I said before, to get this equation I use ideas that start in Gödel's original 1931 paper. But it was Jones and Matijasevič's 1984 paper that finally gave me the tool that I needed.

So that's why I say that there is randomness in elementary number theory, in the arithmetic of the natural numbers. This is an impenetrable stone wall, it's a worst case. From Gödel we knew that we couldn't get a formal axiomatic system to be complete. We knew we were in trouble, and Turing showed us how basic it was, but Ω is an extreme case where reasoning fails completely.

I won't go into the details, but let me talk in vague information-theoretic terms. Matijasevič's equation gives you N arithmetical questions with yes/no answers which turn out to be only $\log N$ bits of algorithmic information. My equation gives you N arithmetical questions with yes/no answers which are irreducible, incompressible mathematical information.

5. Experimental mathematics

Okay, let me say a little bit in the minutes I have left about what this all means.

First of all, the connection with physics. There was a big controversy when quantum mechanics was developed, because quantum theory is nondeterministic. Einstein didn't like that. He said, "God doesn't play dice!" But as I'm sure you all know, with chaos and nonlinear dynamics we've now realized that even in classical physics we get randomness and unpredictability. My work is in the same spirit. It shows that pure mathematics, in fact even elementary number theory, the arithmetic of the natural numbers, 1, 2, 3, 4, 5, is in the same boat. We get randomness there too. So, as a newspaper headline would put it, God not only plays dice in quantum mechanics and in classical physics, but even in pure mathematics, even in elementary number theory. So if a new paradigm is emerging, randomness is at the heart of it. By the way, randomness is also at the heart of quantum field theory, as virtual particles and Feynman path integrals (sums over all histories) show very clearly. So my work fits in with a lot of work in physics, which is why I often get invited to talk at physics meetings.

However the really important question isn't physics, it's mathematics. I've heard that Gödel wrote a letter to his mother who stayed in Europe. You know, Gödel and Einstein were friends at the Institute for Advanced Study. You'd see them walking down the street together. Apparently Gödel wrote a letter to his mother saying that even though Einstein's work on physics had really had a tremendous impact on how people did physics, he was disappointed that his work had not had the same effect on mathematicians. It hadn't made a difference in how mathematicians actually carried on their everyday work. So I think that's the key question: How should you really do mathematics?

I'm claiming I have a much stronger incompleteness result. If so maybe it'll be clearer whether mathematics should be done the ordinary way. What is the ordinary way of doing mathematics? In spite of the fact that everyone knows that any finite set of axioms is incomplete, how do mathematicians actually work? Well suppose you have a conjecture that you've been thinking about for a few weeks, and you believe it because you've tested a large number of cases on a computer. Maybe it's a conjecture about the primes and for two weeks you've tried to prove it. At the end of two weeks you don't say, well obviously the reason I haven't been able to show this is because of Gödel's incompleteness theorem! Let us therefore add it as a new axiom! But if you took Gödel's incompleteness theorem very seriously this might in fact be the way to proceed. Mathematicians will laugh but physicists actually behave this way.

Look at the history of physics. You start with Newtonian physics. You cannot get Maxwell's equations from Newtonian physics. It's a new domain of experience—you need new postulates to deal with it. As for special relativity, well, special relativity is almost in Maxwell's equations. But Schrödinger's equation does not come from Newtonian physics and Maxwell's equations. It's a new domain of experience and again you need new axioms. So physicists are used to the idea that when you start experimenting at a smaller scale, or with new phenomena, you may need new principles to understand and explain what's going on.

Now in spite of incompleteness mathematicians don't behave at all like physicists do. At a subconscious level they still assume that the small number of principles, of postulates and methods of inference, that they learned early as mathematics students, are enough. In their hearts they believe that if you can't prove a result it's your own fault. That's probably a good attitude to take rather than to blame someone else, but let's look at a question like the Riemann hypothesis. A physicist would say that there is ample experimental evidence for the Riemann hypothesis and would go ahead and take it as a working assumption.

What is the Riemann hypothesis? There are many unsolved questions involving the distribution of the prime numbers that can be settled if you assume the Riemann hypothesis. Using computers people check these conjectures and they work beautifully. They're neat formulas but nobody can prove them. A lot of them follow from the Riemann hypothesis. To a physicist this would be enough: It's useful, it explains a lot of data. Of course a physicist then has to be prepared to say "Oh oh, I goofed!" because an experiment can subsequently contradict a theory. This happens very often.

In particle physics you throw up theories all the time and most of them quickly die. But mathematicians don't like to have to backpedal. But if you play it safe, the problem is that you may be losing out, and I believe you are.

I think it should be obvious where I'm leading. I believe that elementary number theory and the rest of mathematics should be pursued more in the spirit of experimental science, and that you should be willing to adopt new principles. I believe that Euclid's statement that an axiom is a self-evident truth is a big mistake. The Schrödinger equation certainly isn't a self-evident truth! And the Riemann hypothesis isn't self-evident either, but it's very useful.

So I believe that we mathematicians shouldn't ignore incompleteness. It's a safe thing to do but we're losing out on results that we could get. It would be as if physicists said, okay no Schrödinger equation, no Maxwell's equations, we stick with Newton, everything must be deduced from Newton's laws. (Maxwell even tried it. He had a mechanical model of an electromagnetic field. Fortunately they don't teach that in college!)

I proposed all this twenty years ago when I started getting these information-theoretic incompleteness results. But independently a new school on the philosophy of mathematics is emerging called the "quasi-empirical" school of thought regarding the foundations of mathematics. There's a book of Tymoczko's called *New Directions in the Philosophy of*

Mathematics (Birkhäuser, Boston, 1986). It's a good collection of articles. Another place to look is *Searching for Certainty* by John Casti (Morrow, New York, 1990) which has a good chapter on mathematics. The last half of the chapter talks about this quasi-empirical view.

By the way, Lakatos, who was one of the people involved in this new movement, happened to be at Cambridge at that time. He'd left Hungary.

The main schools of mathematical philosophy at the beginning of this century were Russell and Whitehead's view that logic was the basis for everything, the formalist school of Hilbert, and an "intuitionist" constructivist school of Brouwer. Some people think that Hilbert believed that mathematics is a meaningless game played with marks of ink on paper. Not so! He just said that to be absolutely clear and precise what mathematics is all about, we have to specify the rules determining whether a proof is correct so precisely that they become mechanical. Nobody who thought that mathematics is meaningless would have been so energetic and done such important work and been such an inspiring leader.

Originally most mathematicians backed Hilbert. Even after Gödel and even more emphatically Turing showed that Hilbert's dream didn't work, in practice mathematicians carried on as before, in Hilbert's spirit. Brouwer's constructivist attitude was mostly considered a nuisance. As for Russell and Whitehead, they had a lot of problems getting all of mathematics from logic. If you get all of mathematics from set theory you discover that it's nice to define the whole numbers in terms of sets (von Neumann worked on this). But then it turns out that there's all kinds of problems with sets. You're not making the natural numbers more solid by basing them on something which is more problematical.

Now everything has gone topsy-turvy. It's gone topsy-turvy, not because of any philosophical argument, not because of Gödel's results or Turing's results or my own incompleteness results. It's gone topsy-turvy for a very simple reason—the computer!

The computer as you all know has changed the way we do everything. The computer has enormously and vastly increased mathematical experience. It's so easy to do calculations, to test many cases, to run experiments on the computer. The computer has so vastly increased mathematical experience, that in order to cope, people are forced to proceed in a more pragmatic fashion. Mathematicians are proceeding more pragmatically, more like experimental scientists do. This new tendency is often called "experimental mathematics." This phrase comes up a lot in the field of chaos, fractals and nonlinear dynamics.

It's often the case that when doing experiments on the computer, numerical experiments with equations, you see that something happens, and you conjecture a result. Of course it's nice if you can prove it. Especially if the proof is short. I'm not sure that a thousand page proof helps too much. But if it's a short proof it's certainly better than not having a proof. And if you have several proofs from different viewpoints, that's very good.

But sometimes you can't find a proof and you can't wait for someone else to find a proof, and you've got to carry on as best you can. So now mathematicians sometimes go ahead with working hypotheses on the basis of the results of computer experiments. Of course if it's physicists doing these computer experiments, then it's certainly okay; they've always relied heavily on experiments. But now even mathematicians sometimes operate in this manner. I believe that there's a new journal called the *Journal of Experimental Mathematics.* They should've put me on their editorial board, because I've been proposing this for twenty years based on my information-theoretic ideas.

So in the end it wasn't Gödel, it wasn't Turing, and it wasn't my results that are making mathematics go in an experimental mathematics direction, in a quasi-empirical direction. The reason that mathematicians are changing their working habits is the computer. I think it's an excellent joke! (It's also funny that of the three old schools of mathematical philosophy, logicist, formalist, and intuitionist, the most neglected was Brouwer, who had a constructivist attitude years before the computer gave a tremendous impulse to constructivism.)

Of course, the mere fact that everybody's doing something doesn't mean that they ought to be. The change in how people are behaving isn't because of Gödel's theorem or Turing's theorems or my theorems, it's because of the computer. But I think that the sequence of work that I've outlined does provide some theoretical justification for what everybody's doing anyway without worrying about the theoretical justification. And I think that the question of how we should actually do mathematics requires **at least** another generation of work. That's basically what I wanted to say—thank you very much!

Bibliography

[1] G. J. Chaitin, *Information-Theoretic Incompleteness,* World Scientific, 1992.

[2] G. J. Chaitin, *Information, Randomness & Incompleteness,* second edition, World Scientific, 1990.

[3] G. J. Chaitin, *Algorithmic Information Theory,* revised third printing, Cambridge University Press, 1990.

A century of controversy over the foundations of mathematics

Lecture given Friday 30 April 1999 at UMass-Lowell. The lecture was video-taped; this is an edited transcript.

I'd like to talk about some crazy stuff. The general idea is that sometimes ideas are very powerful. I'd like to talk about theory, about the computer as a concept, a philosophical concept.

We all know that the computer is a very practical thing out there in the real world! It pays for a lot of our salaries, right? But what people don't remember as much is that really—I'm going to exaggerate, but I'll say it— the computer was invented in order to help to clarify a question about the foundations of mathematics, a philosophical question about the foundations of mathematics.

Now that sounds absurd, but there's some truth in it. There are actually lots of threads that led to the computer, to computer technology, which come from mathematical logic and from philosophical questions about the limits and the power of mathematics.

The computer pioneer Turing was inspired by these questions. Turing was trying to settle a question of Hilbert's having to do with the philosophy of mathematics, when he invented a thing called the Turing machine, which is a mathematical model of a toy computer. Turing did this before there were any real computers, and then he went on to actually build computers. The first computers in England were built by Turing.

And von Neumann, who was instrumental in encouraging the creation of computers as a technology in the United States, (unfortunately as part of a war effort, as part of the effort to build the atom bomb), he knew Turing's work very well. I learned of Turing by reading von Neumann talking about the importance of Turing's work.

So what I said about the origin of the computer isn't a complete lie, but it is a forgotten piece of intellectual history. In fact, let me start off with the final conclusion of this talk... In a way, a lot of this came from work of Hilbert. Hilbert, who was a very well-known German mathematician around the beginning of this century, had proposed formalizing completely all of mathematics, all of mathematical reasoning—deduction. And this proposal of his is a tremendous, glorious failure!

In a way, it's a spectacular failure. Because it turned out that you couldn't formalize mathematical reasoning. That's a famous result of Gödel's that I'll tell you about, done in 1931.

But in another way, Hilbert was really right, because formalism has been the biggest success of this century. Not for reasoning, not for deduction, but for programming, for calculating, for computing, that's where formalism has been a tremendous success. If you look at work by logicians at the beginning of this century, they were talking about formal languages for reasoning and deduction, for doing mathematics and symbolic logic, but they also invented some early versions of programming languages. And **these** are the formalisms that we all live with and work with now all the time! They're a tremendously important technology.

So formalism for reasoning did not work. Mathematicians don't reason in formal languages. But formalism for computing, programming languages, are, in a way, what was right in the formalistic vision that goes back to Hilbert at the beginning of this century, which was intended to clarify epistemological, philosophical questions about mathematics.

So I'm going to tell you this story, which has a very surprising outcome. I'm going to tell you this surprising piece of intellectual history.

The Crisis in Set Theory

So let me start roughly a hundred years ago, with Cantor...

Georg Cantor

The point is this. Normally you think that pure mathematics is static, unchanging, perfect, absolutely correct, absolute truth... Right? Physics may be tentative, but math, things are certain there! Well, it turns out that's not exactly the case.

In this century, in this past century there was a lot of controversy over the foundations of mathematics, and how you should do math, and what's right and what isn't right, and what's a valid proof. Blood was almost shed over this... People had terrible fights and ended up in insane asylums over this. It was a fairly serious controversy. This isn't well known, but I think it's an interesting piece of intellectual history.

More people are aware of the controversy over relativity theory. Einstein was very controversial at first. And then of the controversy over quantum mechanics... These were the two revolutions in the physics of this century. But what's less well known is that there were tremendous revolutions and controversies in pure mathematics too. I'd like to tell you about this. It really all starts in a way from Cantor.

Georg Cantor

What Cantor did was to invent a theory of infinite sets.

Infinite Sets

He did it about a hundred years ago; it's really a little more than a hundred years ago. And it was a tremendously revolutionary theory, it was **extremely** adventurous. Let me tell you why.

Cantor said, let's take 1, 2, 3, ...

$$1, 2, 3, \ldots$$

We've all seen these numbers, right?! And he said, well, let's add an infinite number after this.

$$1, 2, 3, \ldots \omega$$

He called it ω, lowercase Greek omega. And then he said, well, why stop here? Let's go on and keep extending the number series.

$$1, 2, 3, \ldots \omega, \omega + 1, \omega + 2, \ldots$$

Omega plus one, omega plus two, then you go on for an infinite amount of time. And what do you put afterwards? Well, two omega? (Actually, it's omega times two for technical reasons.)

$$1, 2, 3, \ldots \omega \ldots 2\omega$$

Then two omega plus one, two omega plus two, two omega plus three, two omega plus four...

$$1, 2, 3, \ldots 2\omega, 2\omega + 1, 2\omega + 2, 2\omega + 3, 2\omega + 4, \ldots$$

Then you have what? Three omega, four omega, five omega, six omega, ...

$$1, 2, 3, \ldots 3\omega \ldots 4\omega \ldots 5\omega \ldots 6\omega \ldots$$

Well, what will come after all of these? Omega squared! Then you keep going, omega squared plus one, omega squared plus six omega plus eight... Okay, you keep going for a long time, and the next interesting thing after omega squared will be? Omega cubed! And then you have omega to the fourth, omega to the fifth, and much later?

$$1, 2, 3, \ldots \omega \ldots \omega^2 \ldots \omega^3 \ldots \omega^4 \ldots \omega^5$$

Omega to the omega!
$$1, 2, 3, \ldots \omega \ldots \omega^2 \ldots \omega^\omega$$

And then much later it's omega to the omega to the omega an infinite number of times!
$$1, 2, 3, \ldots \omega \ldots \omega^2 \ldots \omega^\omega \ldots \omega^{\omega^{\omega^{\omega^{\cdots}}}}$$

I think this is usually called epsilon nought.

$$\varepsilon_0 = \omega^{\omega^{\omega^{\omega^{\cdots}}}}$$

It's a pretty mind-boggling number! After this point things get a little complicated...

And this was just one little thing that Cantor did as a warm-up exercise for his main stuff, which was measuring the size of infinite sets! It was spectacularly imaginative, and the reactions were extreme. Some people loved what Cantor was doing, and some people thought that he should be put in an insane asylum! In fact he had a nervous breakdown as a result of those criticisms. Cantor's work was very influential, leading to point-set topology and other abstract fields in the mathematics of the twentieth century. But it was also very controversial. Some people said, it's theology, it's not real, it's a fantasy world, it has nothing to do with serious math! And Cantor never got a good position and he spent his entire life at a second-rate institution.

Bertrand Russell's Logical Paradoxes

Then things got even worse, due mainly, I think, to Bertrand Russell, one of my childhood heroes.

Bertrand Russell

Bertrand Russell was a British philosopher who wrote beautiful essays, very individualistic essays, and I think he got the Nobel prize in literature for his wonderful essays. Bertrand Russell started off as a mathematician and then degenerated into a philosopher and finally into a humanist; he went downhill rapidly! Anyway, Bertrand Russell discovered a whole bunch of disturbing paradoxes, first in Cantor's theory, then in logic itself. He found cases where reasoning that seemed to be okay led to contradictions.

And I think that Bertrand Russell was tremendously influential in spreading the idea that there was a serious crisis and that these contradictions had to be resolved somehow. The paradoxes that Russell discovered attracted a great deal of attention, but strangely enough only one of them ended up with Russell's name on it! For example, one of these paradoxes is called the Burali-Forti paradox, because when Russell published it he stated in a footnote that it had been suggested to him by reading a paper by Burali-Forti. But if you look at the paper by Burali-Forti, you don't see the paradox!

But I think that the realization that something was seriously wrong, that something was rotten in the state of Denmark, that reasoning was bankrupt and something had to be done about it pronto, is due principally to Russell. Alejandro Garciadiego, a Mexican historian of math, has written a book which suggests that Bertrand Russell really played a much bigger role in this than is usually realized: Russell played a key role in formulating not only the Russell paradox, which bears his name, but also the Burali-Forti paradox and the Berry paradox, which don't. Russell was instrumental in discovering them and in realizing their significance. He told everyone that they were important, that they were not just childish word-play.

Anyway, the best known of these paradoxes is called the Russell paradox nowadays. You consider the set of all sets that are not members of themselves. And then you ask, "Is this set a member of itself or not?" If it is a member of itself, then it shouldn't be, and vice versa! It's like the barber in a small, remote town who shaves all the men in the town who don't shave themselves. That seems pretty reasonable, until you ask "Does the barber shave himself?" He shaves himself if and only if he doesn't shave himself, so he can't apply that rule to himself!

Now you may say, "Who cares about this barber!" It was a silly rule anyway, and there are always exceptions to the rule! But when you're dealing with a **set**, with a mathematical concept, it's not so easy to dismiss the problem. Then it's not so easy to shrug when reasoning that seems to be okay gets you into trouble!

By the way, the Russell paradox is a set-theoretic echo of an earlier paradox, one that was known to the ancient Greeks and is called the Epimenides paradox by some philosophers. That's the paradox of the liar: "This statement is false!" "What I'm now saying is false, it's a lie." Well, is it false? If it's false, if something is false, then it doesn't correspond with reality. So if I'm saying this statement is false, that means that it's not false—which means that it must be true. But if it's true, and I'm saying it's false, then it must be false! So whatever you do you're in trouble!

So you can't get a definite logical truth value, everything flip flops, it's neither true nor false. And you might dismiss this and say that these are just meaningless word games, that it's not serious. But Kurt Gödel later built his work on these paradoxes, and he had a very different opinion.

Kurt Gödel

He said that Bertrand Russell made the amazing discovery that our logical intuitions, our mathematical intuitions, are self-contradictory, they're inconsistent! So Gödel took Russell very seriously, he didn't think that it was all a big joke.

Now I'd like to move on and tell you about David Hilbert's rescue plan for dealing with the crisis provoked by Cantor's set theory and by Russell's paradoxes.

David Hilbert

David Hilbert to the Rescue with Formal Axiomatic Theories

One of the reactions to the crisis provoked by Cantor's theory of infinite sets, one of the reactions was, well, let's escape into **formalism**. If we get into trouble with reasoning that seems okay, then one solution is to use symbolic logic, to create an artificial language where we're going to be very careful and say what the rules of the game are, and make sure that we don't get the contradictions. Right? Because here's a piece of reasoning that looks okay but it leads to a contradiction. Well, we'd like to get rid of that. But natural language is ambiguous—you never know what a pronoun refers to. So let's create an artificial language and make things very, very precise and make sure that we get rid of all the contradictions! So this was the notion of formalism.

Formalism

Now I don't think that Hilbert actually intended that mathematicians should work in such a perfect artificial language. It would sort of be like a programming language, but for reasoning, for doing mathematics, for deduction, not for computing, that was Hilbert's idea. But he never expressed it that way, because there were no programming languages back then.

So what are the ideas here? First of all, Hilbert stressed the importance of the axiomatic method.

Axiomatic Method

The notion of doing mathematics that way goes back to the ancient Greeks and particularly to Euclidean geometry, which is a beautifully clear mathematical system. But that's not enough; Hilbert was also saying that we should use symbolic logic.

Symbolic Logic

And symbolic logic also has a long history: Leibniz, Boole, Frege, Peano... These mathematicians wanted to make reasoning like algebra. Here's how Leibniz put it: He talked about avoiding disputes—and he was probably thinking of political disputes and religious disputes—by calculating who was right instead of arguing about it! Instead of fighting, you should be able to sit down at a table and say, "Gentleman, let us compute!" What a beautiful fantasy!...

So the idea was that mathematical logic should be like arithmetic and you should be able to just grind out a conclusion, no uncertainty, no questions of interpretation. By using an artificial math language with a symbolic logic you should be able to achieve *perfect rigor*. You've heard the word "rigor", as in "rigor mortis", used in mathematics? It's not that rigor! But the idea is that an argument is either completely correct or else it's total nonsense, with nothing in between. And a proof that is formulated in a formal axiomatic system should be absolutely clear, it should be completely sharp!

In other words, Hilbert's idea was that we should be completely precise about what the rules of the game are, and about the definitions, the elementary concepts, and the grammar and the language—all the rules of the game—so that we can all agree on how mathematics should be done. In practice it would be too much work to use such a formal axiomatic system, but it would be philosophically significant because it would settle once and for all the question of whether a piece of mathematical reasoning is correct or incorrect.

Okay? So Hilbert's idea seemed fairly straightforward. He was just following the axiomatic and the formal traditions in mathematics. Formal as in formalism, as in using formulas, as in calculating! He wanted to go all the way, to the very end, and formalize all of mathematics, but it seemed like a fairly reasonable plan. Hilbert wasn't a revolutionary, he was a conservative... The amazing thing, as I said before, was that it turned out that Hilbert's rescue plan **could not work**, that it couldn't be done, that it was **impossible** to make it work!

Hilbert was just following the whole mathematics tradition up to that point: the axiomatic method, symbolic logic, formalism... He wanted to avoid the paradoxes by being absolutely precise, by creating a completely formal axiomatic system, an artificial language, that avoided the paradoxes, that made them impossible, that **outlawed** them! And most mathematicians probably thought that Hilbert was right, that *of course* you could do this—it's just the notion that in mathematics things are absolutely clear, black or white, true or false.

So Hilbert's idea was just an extreme, an exaggerated version of the normal notion of what mathematics is all about: the idea that we can decide and agree on the rules of the game, all of them, once and for all. The big surprise is that it turned out that this **could not** be done. Hilbert turned out to be wrong, but wrong in a tremendously fruitful way, because he had asked a very good question. In fact, by asking this question he actually created an entirely new field of mathematics called **meta**mathematics.

Metamathematics

Metamathematics is mathematics turned inward, it's an introspective field of math in which you study what mathematics can achieve or can't achieve.

What is Metamathematics?

That's my field—metamathematics! In it you look at mathematics from above, and you use mathematical reasoning to discuss what mathematical reasoning can or cannot achieve. The basic idea is this: Once you entomb mathematics in an artificial language *à la* Hilbert, once you set up a completely formal axiomatic system, then you can forget that it has any meaning and just look at it as a game that you play with marks on paper that enables you to deduce theorems from axioms. You can forget about the meaning of this game, the game of mathematical reasoning, it's just combinatorial play with symbols! There are certain rules, and you can study these rules and forget that they have any meaning!

What things do you look at when you study a formal axiomatic system from above, from the outside? What kind of questions do you ask?

Well, one question you can ask is if you can prove that "0 equals 1"?

$$0 = 1 ?$$

Hopefully you can't, but how can you be sure? It's hard to be sure!

And for any question A, for any affirmation A, you can ask if it's possible to settle the matter by either proving A or the opposite of A, not A.

$$A ? \ \neg A ?$$

That's called *completeness.*

Completeness

A formal axiomatic system is complete if you can settle any question A, either by proving it (A), or by proving that it's false $(\neg A)$. That would be nice! Another interesting question is if you can prove an assertion (A) and you can also prove the contrary assertion $(\neg A)$. That's called *inconsistency,* and if that happens it's very bad! *Consistency* is much better than *inconsistency*!

Consistency

So what Hilbert did was to have the remarkable idea of creating a new field of mathematics whose subject would be mathematics itself. But you can't do this until you have a completely formal axiomatic system. Because as long as any "meaning" is involved in mathematical reasoning, it's all subjective. Of course, the reason we do mathematics is because it has meaning, right? But if you want to be able to study mathematics, the power of mathematics, using mathematical methods, you have to "desiccate" it to "crystallize out" the meaning and just be left with an artificial language with completely precise rules, in fact, with one that has a **mechanical proof-checking algorithm**.

Proof-Checking Algorithm

The key idea that Hilbert had was to envision this perfectly desiccated or crystallized axiomatic system for all of mathematics, in which the rules would be so precise that if someone had a proof there would be a referee, there would be a mechanical procedure, which would either say "This proof obeys the rules" or "This proof is wrong; it's breaking the rules". That's how you get the criterion for mathematical truth to be completely objective and not to depend on meaning or subjective understanding: by reducing it all to calculation. Somebody says "This is a proof", and instead of having to submit it to a human referee who takes two years to decide if the paper is correct, instead you just give it to a machine. And the machine eventually says "This obeys the rules" or "On line 4 there's a misspelling" or "This thing on line 4 that supposedly follows from line 3, actually doesn't". And that would be the end, no appeal!

The idea was not that mathematics should actually be done this way. I think that that's calumny, that's a false accusation. I don't think that Hilbert really wanted to turn mathematicians into machines. But the idea was that if you could take mathematics and do it this way, then you could use mathematics to study the power of mathematics. And **that** is the important new thing that Hilbert came up with. Hilbert wanted to do this in order to reaffirm the traditional view of mathematics, in order to justify himself. . .

He proposed having one set of axioms and this formal language, this formal system, which would include all of mathematical reasoning, that we could all agree on, and that would be **perfect**! We'd then know all the rules of the game. And he just wanted to use metamathematics to show that this formal axiomatic system was good—that it was consistent and that it was complete—in order to convince people to accept it. This would have settled once and for all the philosophical questions "When is a proof correct?" and "What is mathematical truth?" Like this everyone could agree on whether a mathematical proof is correct or not. And in fact we used to think that this was an objective thing.

In other words, Hilbert's just saying, if it's really objective, if there's no subjective element, and a mathematical proof is either true or false, well, then there should be certain rules for deciding that and it shouldn't depend, if you fill in all the details, it shouldn't depend on interpretation. It's important to fill in all the details—that's the idea of mathematical logic, to "atomize" mathematical reasoning into such tiny steps that **nothing** is left to the imagination, nothing is left out! And if nothing is left out, then a proof can be checked automatically, that was Hilbert's point, that's really what symbolic logic is all about.

And Hilbert thought that he was actually going to be able to do this. He was going to formalize all of mathematics, and we were all going to agree that these were in fact the rules of the game. Then there'd be just one version of mathematical truth, not many variations. We don't want to have a German mathematics and a French mathematics and a Swedish mathematics and an American mathematics, no, we want a **universal** mathematics, one universal criterion for mathematical truth! Then a paper that is done by a mathematician in one country can be understood by a mathematician in another country. Doesn't that sound reasonable?! So you can imagine just how very, very shocking it was in 1931 when Kurt Gödel showed that it wasn't at all reasonable, that it could **never** be done!

1931 Kurt Gödel

Kurt Gödel Discovers Incompleteness

Gödel did this is Vienna, but he was from what I think is now called the Czech republic, from the city of Brünn or Brno. It was part of the Austro-Hungarian empire then, but now it's a separate country. And later he was at the Institute for Advanced Study in Princeton, where I visited his grave a few weeks ago. And the current owner of Gödel's house was nice enough to invite me in when he saw me examining the house, instead of calling the police! They know they're in a house that some people are interested in for historical reasons.

Okay, so what did Kurt Gödel do? Well, Gödel sort of exploded this whole view of what mathematics is all about. He came up with a famous incompleteness result, "Gödel's incompleteness theorem".

Incompleteness

And there's a lovely book explaining the way Gödel originally did it. It's by Nagel and Newman, and it's called *Gödel's Proof.* I read it when I was a child, and forty years later it's still in print!

What is this amazing result of Gödel's? Gödel's amazing discovery is that Hilbert was **wrong**, that it cannot be done, that there's **no way** to take all of mathematical truth and to agree on a set of rules and to have a formal axiomatic system for all of mathematics in which it is crystal clear whether something is correct or not!

More precisely, what Gödel discovered was that if you just try to deal with elementary arithmetic, with 0, 1, 2, 3, 4... and with addition and multiplication

$$+ \times 0, 1, 2, 3, 4, \ldots$$

—this is "elementary number theory" or "arithmetic"—and you just try to have a set of axioms for this—the usual axioms are called Peano arithmetic— even this can't be done! Any set of axioms that tries to have **the whole truth** and **nothing but the truth** about addition, multiplication, and 0, 1, 2, 3, 4, 5, 6, 7, 8, 9, 10... will have to be incomplete. More precisely, it'll either be inconsistent or it'll be incomplete. So if you assume that it only tells the truth, then it won't tell the whole truth. There's no way to capture all the truth about addition, multiplication, and 0, 1, 2, 3, 4...! In particular, if you assume that the axioms don't allow you to prove false theorems, then it'll be incomplete, there'll be true theorems that you cannot prove from these axioms!

This is an absolutely devastating result, and all of traditional mathematical philosophy ends up in a heap on the floor! At the time this was considered to be absolutely devastating. However you may notice that in 1931 there were also a few other problems to worry about. The situation in Europe was bad. There was a major depression, and a war was brewing. I agree, not all problems are mathematical! There's more to life than epistemology! But you begin to wonder, well, if the traditional view of mathematics isn't correct, then **what is correct?** Gödel's incompleteness theorem was very surprising and a terrible shock.

How did Gödel do it? Well, Gödel's proof is very clever. It almost looks crazy, it's very paradoxical. Gödel starts with the paradox of the liar, "I'm false!", which is neither true nor false.

"This statement is false!"

And what Gödel does is to construct a statement that says of itself "I'm unprovable!"

                    "This statement is unprovable!"

Now if you can construct such a statement in elementary number theory, in arithmetic, a mathematical statement—I don't know how you make a mathematical statement say it's unprovable, you've got to be very clever— but if you can do it, it's easy to see that you're in trouble. Just think about it a little bit. It's easy to see that you're in trouble. Because if it's provable, it's false, right? So you're in trouble, you're proving false results. And if it's unprovable and it says that it's unprovable, then it's true, and mathematics is incomplete. So either way, you're in trouble! Big trouble!

And Gödel's original proof is very, very clever and hard to understand. There are a lot of complicated technical details. But if you look at his original paper, it seems to me that there's a lot of LISP programming in it, or at least something that looks a lot like LISP programming. Anyway, now we'd call it LISP programming. Gödel's proof involves defining a great many functions recursively, and these are functions dealing with lists, which is precisely what LISP is all about. So even though there were no programming languages in 1931, with the benefit of hindsight you can clearly see a programming language in Gödel's original paper. And the programming language I know that's closest to it is LISP, pure LISP, LISP without side-effects, interestingly enough—that's the heart of LISP.

So this was a very, very shocking result, and people didn't really know what to make of it.

Now the next major step forward comes only five years later, in 1936, and it's by Alan Turing.

                    1936 Alan Turing

Alan Turing Discovers Uncomputability

Turing's approach to all these questions is completely different from Gödel's, and much deeper. Because Turing brings it out of the closet! What he brings out of the closet is the computer! The computer was implicit in Gödel's paper, but this was really not visible to any ordinary mortal, not at that time, only with hindsight. And Turing really brings it out in the open.

Hilbert had said that there should be a "mechanical procedure" to decide if a proof obeys the rules or not. And Hilbert never clarified what he meant by a mechanical procedure, it was all words. But, Turing said, what you really mean is a machine, and a machine of a kind that we now call a Turing machine—but it wasn't called that in Turing's original paper. In fact, Turing's original paper contains a programming language, just like Gödel's paper does, what we would now call a programming language. But the two programming languages are very different. Turing's programming language isn't a high-level language like LISP, it's more like a machine language. In fact, it's a horrible machine language, one that nobody would want to use today, because it's too simple.

But Turing makes the point that even though Turing machines are very simple, even though their machine language is rather primitive, they're very flexible, very general-purpose machines. In fact, he claims, any computation that a human being can perform, should be **possible** to do using such a machine. Turing's train of thought now takes a very dramatic turn. What, he asks, is **impossible** for such a machine? What can't it do? And he immediately finds a question that no Turing machine can settle, a problem that no Turing machine can solve. That's *the halting problem,* the problem of deciding in advance if a Turing machine or a computer program will eventually halt.

The Halting Problem

So the shocking thing about this 1936 paper is that first of all he comes up with the notion of a general-purpose or universal computer, with a machine that's flexible, that can do what any machine can do. One calculating machine that can do **any** calculation, which is, we now say, a general-purpose computer. And then he immediately shows that there are limits to what such a machine can do. And how does he find something that cannot be done by any such machine? Well, it's very simple! It's the question of whether a computer program will eventually halt, with no time limit.

If you put a time limit, it's very easy. If you want to know if a program halts in a year, you just run it for a year, and either it halted or doesn't. What Turing showed is that you get in terrible trouble if there's no time limit. Now you may say, "What good is a computer program that takes more than a year, that takes more than a thousand years?! There's always a time limit!" I agree, this is pure math, this is not the real world. You only get in trouble with **infinity**! But Turing shows that if you put no time limit, then you're in real difficulties.

So this is called *the halting problem.* And what Turing showed is that there's no way to decide in advance if a program will eventually halt.

The Halting Problem

If it **does** halt, by running it you can eventually discover that, if you're just patient. The problem is you don't know when to give up. And Turing was able to show with a very simple argument which is just Cantor's diagonal argument—coming from Cantor's theory of infinite sets, by the way—I don't have time to explain all this—with a very simple argument Turing was able to show that this problem

The Halting Problem

cannot be solved.

No computer program can tell you in advance if another computer program will eventually halt or not. And the problem is the ones that don't halt, that's really the problem. The problem is knowing when to give up.

So now the interesting thing about this is that Turing immediately deduces as a corollary that if there's no way to decide in advance **by a calculation** if a program will halt or not, well then there cannot be any way to **deduce** it in advance using reasoning either. No formal axiomatic system can enable you to deduce in advance whether a program will halt or not.

Because if you can use a formal axiomatic system to always deduce whether a program will halt or not, well then, that will give you a way to calculate in advance whether a program will halt or not. You simply run through all possible deductions—you can't do this in practice—but in principle you can run through all possible proofs in size order, checking which ones are correct, until either you find a proof that the program will halt eventually or you find a proof that it's never going to halt.

This is using the idea of a completely formal axiomatic system where you don't need a mathematician—you just run through this calculation on a computer—it's mechanical to check if a proof is correct or not. So if there were a formal axiomatic system which always would enable you to prove, to deduce, whether a program will halt or not, that would give you a way to calculate in advance whether a program will halt or not. And that's impossible, because you get into a paradox like "This statement is false!" You get a program that halts if and only if it doesn't halt, that's basically the problem. You use an argument having the same flavor as the Russell paradox.

So Turing went more deeply into these questions than Gödel. As a student I read Gödel's proof, and I could follow it step by step: I read it in Nagel and Newman's book, which is a lovely book. It's a marvelous book, it's so understandable! It's still in print, and it was published in 1958...But I couldn't really feel that I was coming to grips with Gödel's proof, that I could really understand it. The whole thing seemed too delicate, it seemed too fragile, it seemed too superficial...And there's this business in the closet about computing, that's there in Gödel, but it's hidden, it's not in the open, we're not really coming to terms with it.

Now Turing is really going, I think, much deeper into this whole matter. And he's showing, by the way, that it's not just one particular axiomatic system, the one that Gödel studied, that can't work, but that **no** formal axiomatic system can work. But it's in a slightly different context. Gödel was really looking at 0, 1, 2, 3, 4...and addition and multiplication, and Turing is looking at a rather strange mathematical question, which is does a program halt or not. It's a mathematical question **that did not exist** at the time of Gödel's original paper. So you see, Turing worked with completely new concepts...

But Gödel's paper is not only tremendously clever, he had to have the courage to imagine that Hilbert might be wrong. There's another famous mathematician of that time, von Neumann—whose grave I found near Gödel's, by the way, at Princeton. Von Neumann was probably as clever as Gödel or anyone else, but it never occurred to him that Hilbert could be wrong. And the moment that he heard Gödel explain his result, von Neumann immediately appreciated it and immediately started deducing consequences. But von Neumann said, "I missed it, I missed the boat, I didn't get it right!" And Gödel did, so he was much more profound...

Now Turing's paper is also full of technical details, like Gödel's paper, because there is a programming language in Turing's paper, and Turing also gives a rather large program, which of course has bugs, because he wasn't able to run it and debug it—it's the program for a universal Turing machine. But the basic thing is the ideas, and the new ideas in Turing's work are just breathtaking! So I think that Turing went beyond Gödel, but you have to recognize that Gödel took the first step, and the first step is historically the most difficult one and takes the most courage. To imagine that Hilbert could be wrong, which never occurred to von Neumann, that was something!

I Discover Randomness in Pure Mathematics

Okay, so then what happened? Then World War II begins. Turing starts working on cryptography, von Neumann starts working on how to calculate atom bomb detonations, and people forget about incompleteness for a while.

This is where I show up on the scene. The generation of mathematicians who were concerned with these questions basically passes from the scene with World War II. And I'm a kid in the 1950s in the United States reading the original article by Nagel and Newman in *Scientific American* in 1956 that became their book.

And I didn't realize that mathematicians really preferred to forget about Gödel and go on working on their favorite problems. I'm fascinated by incompleteness and I want to understand it. Gödel's incompleteness result fascinates me, but I can't really understand it, I think there's something fishy... As for Turing's approach, I think it goes much deeper, but I'm still not satisfied, I want to understand it better.

And I get a funny idea about randomness... I was reading a lot of discussions of another famous intellectual issue when I was a kid—not the question of the foundations of mathematics, the question of the foundations of **physics**! These were discussions about relativity theory and cosmology and even more often about quantum mechanics, about what happens in the atom. It seems that when things are very small the physical world behaves in a completely crazy way that is totally unlike how objects behave here in this classroom. In fact things are **random**—intrinsically unpredictable—in the atom.

Einstein hated this. Einstein said that "God doesn't play dice!" By the way, Einstein and Gödel were friends at Princeton, and they didn't talk very much with anybody else, and I heard someone say that Einstein had brainwashed Gödel against quantum mechanics! It was the physicist John Wheeler, who told me that he once asked Gödel if there could be any connection between quantum uncertainty and Gödel's incompleteness theorem, but Gödel refused to discuss it...

Okay, so I was reading about all of this, and I began to wonder—in the back of my head I began to ask myself—could it be that there was also randomness in pure mathematics?

The idea in quantum mechanics is that randomness is fundamental, it's a basic part of the universe. In normal, everyday life we know that things are unpredictable, but in theory, in Newtonian physics and even in Einstein's relativity theory—that's all called *classical* as opposed to *quantum* physics—in theory in classical physics you can predict the future. The equations are *deterministic,* not *probabilistic.* If you know the initial conditions exactly, with infinite precision, you apply the equations and you can predict with infinite precision any future time and even in the past, because the equations work either way, in either direction. The equations don't care about the direction of time...

This is that wonderful thing sometimes referred to as *Laplacian determinism.* I think that it's called that because of Laplace's *Essai Philosophique sur les Probabilités,* a book that was published almost two centuries ago. At the beginning of this book Laplace explains that by applying Newton's laws, in principle a demon could predict the future arbitrarily far, or the past arbitrarily far, if it knew the exact conditions at the current moment. This is not the type of world where you talk about free will and moral responsibility, but if you're doing physics calculations it's a great world, because you can calculate everything!

But in the 1920s with quantum mechanics it began to look like God plays dice in the atom, because the basic equation of quantum mechanics is the Schrödinger equation, and the Schrödinger equation is an equation that talks about the **probability** that an electron will do something. The basic quantity is a probability and it's a wave equation saying how a probability wave interferes with itself. So it's a completely different kind of equation, because in Newtonian physics you can calculate the precise trajectory of a particle and know exactly how it's going to behave. But in quantum mechanics the fundamental equation is an equation dealing with probabilities! That's it, that's all there is!

You **can't know** exactly where an electron is and what its velocity vector is—exactly what direction and how fast it's going. It doesn't have a specific state that's known with infinite precision the way it is in classical physics. If you know very accurately where an electron is, then its velocity—its momentum—turns out to be wildly uncertain. And if you know exactly in which direction and at what speed it's going, then its position becomes infinitely uncertain. That's the infamous *Heisenberg uncertainty principle,* there's a trade-off, that seems to be the way the physical universe works...

It's an interesting historical fact that before people used to hate this—Einstein hated it—but now people think that they can **use** it! There's a crazy new field called *quantum computing* where the idea is to stop fighting it. If you can't lick them, join them! The idea is that maybe you can make a brand new technology using something called *quantum parallelism.* If a quantum computer is uncertain, maybe you can have it uncertainly do many computations at the same time! So instead of fighting it, the idea is to use it, which is a great idea.

But when I was a kid people were still arguing over this. Even though he had helped to create quantum mechanics, Einstein was still fighting it, and people were saying, "Poor guy, he's obviously past his prime!"

Okay, so I began to think that maybe there's also randomness in pure mathematics. I began to suspect that maybe that's **the real reason for incompleteness**. A case in point is elementary number theory, where there are some very difficult questions. Take a look at the prime numbers. Individual prime numbers behave in a very unpredictable way, if you're interested in their detailed structure. It's true that there are **statistical** patterns. There's a thing called *the prime number theorem* that predicts fairly accurately the over-all average distribution of the primes. But as for the detailed distribution of individual prime numbers, that looks pretty random!

So I began to think about **randomness**... I began to think that maybe that's what's really going on, maybe that's a deeper reason for all this incompleteness. So in the 1960s I, and independently some other people, came up with some new ideas. And I like to call this new set of ideas *algorithmic information theory*.

Algorithmic Information Theory

That name makes it sound very impressive, but the basic idea is just to look at the size of computer programs. You see, it's just a *complexity measure,* it's just a kind of *computational complexity...*

I think that one of the first places that I heard about the idea of computational complexity was from von Neumann. Turing came up with the idea of a computer as a mathematical concept—it's a perfect computer, one that never makes mistakes, one that has as much time and space as it needs to work—it's always finite, but the calculation can go on as long as it has to. After Turing comes up with this idea, the next logical step for a mathematician is to study the time, the work needed to do a calculation—its complexity. And in fact I think that around 1950 von Neumann suggested somewhere that there should be a new field which looks at the **time** complexity of computations, and that's now a very well-developed field. So of course if most people are doing that, then I'm going to try something else!

My idea was not to look at the **time**, even though from a practical point of view time is very important. My idea was to look at the **size** of computer programs, at the amount of information that you have to give a computer to get it to perform a given task. From a practical point of view, the amount of information required isn't as interesting as the running time, because of course it's very important for computers to do things as fast as possible...But it turns out that from a conceptual point of view, it's not that way at all. I believe that from a fundamental philosophical point of view, the right question is to look at the **size** of computer programs, not at the **time**. Why?—Besides the fact that it's my idea so obviously I'm going to be prejudiced! The reason is because program-size complexity connects with a lot of fundamental stuff in physics.

You see, in physics there's a notion called *entropy,* which is how disordered a system is. Entropy played a particularly crucial role in the work of the famous 19th century physicist Boltzmann,

Ludwig Boltzmann

and it comes up in the field of statistical mechanics and in thermodynamics. Entropy measures how disordered, how chaotic, a physical system is. A crystal has low entropy, and a gas at high temperature has high entropy. It's the amount of chaos or disorder, and it's a notion of randomness that physicists like.

And entropy is connected with some fundamental philosophical questions—it's connected with the question of the *arrow of time,* which is **another** famous controversy. When Boltzmann invented this wonderful thing called statistical mechanics—his theory is now considered to be one of the masterpieces of 19th century physics, and all physics is now statistical physics—he ended up by committing suicide, because people said that his theory was **obviously** wrong! Why was it obviously wrong? Because in Boltzmann's theory entropy has got to increase and so there's an arrow of time. But if you look at the equations of Newtonian physics, they're time reversible. There's no difference between predicting the future and predicting the past. If you know at one instant exactly how everything is, you can go in either direction, the equations don't care, there's no direction of time, backward is the same as forward.

But in everyday life and in Boltzmann statistical mechanics, there is a difference between going backward and forward. Glasses break, but they don't reassemble spontaneously! And in Boltzmann's theory entropy has got to increase, the system has to get more and more disordered. But people said, "You can't deduce that from Newtonian physics!" Boltzmann was pretending to. He was looking at a gas. The atoms of a gas bounce around like billiard balls, it's a billiard ball model of how a gas works. And each interaction is reversible. If you run the movie backwards, it looks the same. If you look at a small portion of a gas for a small amount of time, you can't tell whether you're seeing the movie in the right direction or the wrong direction.

But Boltzmann gas theory says that there is an arrow of time—a system will start off in an ordered state and will end up in a very mixed up disordered state. There's even a scary expression in German, *heat death.* People said that according to Boltzmann's theory the universe is going to end up in a horrible ugly state of maximum entropy or heat death! This was the dire prediction! So there was a lot of controversy about his theory, and maybe that was one of the reasons that Boltzmann killed himself.

And there is a connection between my ideas and Boltzmann's, because looking at the size of computer programs is very similar to this notion of the degree of disorder of a physical system. A gas takes a large program to say where all its atoms are, but a crystal doesn't take as big a program, because of its regular structure. Entropy and program-size complexity are closely related...

This idea of program-size complexity is also connected with the philosophy of the scientific method. You've heard of *Occam's razor,* of the idea that the simplest theory is best? Well, what's a theory? It's a computer program for predicting observations. And the idea that the simplest theory is best translates into saying that a **concise** computer program is the best theory. What if there is no concise theory, what if the most concise program or the best theory for reproducing a given set of experimental data is **the same size** as the data? Then the theory is no good, it's cooked up, and the data is incomprehensible, it's random. In that case the theory isn't doing a useful job. A theory is good to the extent that it compresses the data into a much smaller set of theoretical assumptions. The greater the compression, the better!—That's the idea...

So this idea of program size has a lot of philosophical resonances, and you can define randomness or maximum entropy as something that cannot be compressed at all. It's an object with the property that basically the only way you can describe it to someone is to say "this is it" and show it to them. Because it has no structure or pattern, there is no concise description, and the thing has to be understood as "a thing in itself", it's irreducible.

Randomness = Incompressibility

The other extreme is an object that has a very regular pattern so you can just say that it's "a million 0s" or "half a million repetitions of 01", pairs 01, 01, 01 repeated half a million times. These are very long objects with a very concise description. Another long object with a concise description is an *ephemeris,* I think it's called that, it's a table giving the positions of the planets as seen in sky, daily, for a year. You can compress all this astronomical information into a small FORTRAN program that uses Newtonian physics to calculate where the planets will be seen in the sky every night.

But if you look at how a roulette wheel behaves, then there is no pattern, the series of outcomes cannot be compressed. Because if there were a pattern, then people could use it to win, and having a casino wouldn't be such a good business! The fact that casinos make lots of money shows that there is no way to predict what a roulette wheel will do, there is no pattern—the casinos make it their job to ensure that!

So I had this new idea, which was to use program-size complexity to define randomness. And when you start looking at the size of computer programs— when you begin to think about this notion of program-size or information complexity instead of run-time complexity—then the interesting thing that happens is that everywhere you turn you immediately find incompleteness! You immediately find things that escape the power of mathematical reasoning, things that escape the power of any computer program. It turns out that they're everywhere!

It's very dramatic! In only three steps we went from Gödel, where it's very surprising that there are limits to reasoning, to Turing, where it looks much more natural, and then when you start looking at program size, well, incompleteness, the limits of mathematics, it just hits you in the face! Why?! Well, **the very first question** that you ask in my theory gets you into trouble. What's that? Well, in my theory I measure the complexity of something by the size of the smallest computer program for calculating it. But how can I be sure that I have the smallest computer program?

Let's say that I have a particular calculation, a particular output, that I'm interested in, and that I have this nice, small computer program that calculates it, and I think that it's the smallest possible program, the most concise one that produces this output. Maybe a few friends of mine and I were trying to do it, and this was the best program that we came up with; nobody did any better. But how can you be **sure**? Well, the answer is that **you can't be sure**. It turns out you **can never be sure!** You can **never** be sure that a computer program is what I like to call *elegant,* namely that it's the most concise one that produces the output that it produces. **Never ever!** This escapes the power of mathematical reasoning, amazingly enough.

But for any computational task, once you fix the computer programming language, once you decide on the computer programming language, and if you have in mind a particular output, there's got to be at least one program that is the smallest possible. There may be a tie, there may be several, right?, but there's got to be at least **one** that's smaller than all the others. But you can never be sure that you've found it!

And the precise result, which is one of my favorite incompleteness results, is that if you have N bits of axioms, you can never prove that a program is elegant—smallest possible—if the program is more than N bits long. That's basically how it works. So any given set of mathematical axioms, any formal axiomatic system in Hilbert's style, can only prove that **finitely many** programs are elegant, are the most concise possible for their output.

To be more precise, you get into trouble with an elegant program if it's larger than a computerized version of the axioms—It's really the size of the proof-checking program for your axioms. In fact, it's the size of the program that runs through all possible proofs producing all possible theorems. If you have in mind a particular programming language, and you need a program of a certain size to implement a formal axiomatic system, that is to say, to write the proof-checking algorithm and to write the program that runs through all possible proofs filtering out all the theorems, if that program is a certain size in a language, and if you look at programs in that same language that are larger, then you can never be sure that such a program is elegant, you can never prove that such a program is elegant using the axioms that are implemented in the same language by a smaller program. That's basically how it works.

So there are an infinity of elegant programs out there. For any computational task there's got to be at least one elegant program, and there may be several, but you can never be sure except in a finite number of cases. That's my result, and I'm very proud of it!

So it turns out that you can't calculate the program-size complexity, you can never be sure what the program-size complexity of anything is. Because to determine the program-size complexity of something is to know the size of the most concise program that calculates it—but that means—it's essentially the same problem—then I would know that this program is the most concise possible, I would know that it's an elegant program, and you can't do that if the program is larger than the axioms. So if it's N bits of axioms, you can never determine the program-size complexity of anything that has more than N bits of complexity, which means almost everything, because almost everything has more than N bits of complexity. Almost everything has more complexity than the axioms that you're using.

Why do I say that? The reason for using axioms is because they're simple and believable. So the sets of axioms that mathematicians normally use are fairly concise, otherwise no one would believe in them! Which means that in practice there's this vast world of mathematical truth out there, which is an infinite amount of information, but any given set of axioms only captures a tiny finite amount of this information! And that's why we're in trouble, that's my bottom line, that's my final conclusion, that's the real dilemma.

So in summary, I have two ways to explain why I think Gödel incompleteness is natural and inevitable rather than mysterious and surprising. The two ways are—that the idea of randomness in physics, that some things make no sense, also happens in pure mathematics, is one way to say it. But a better way to say it, is that mathematical truth is an infinite amount of information, but any particular set of axioms just has a finite amount of information, because there are only going to be a finite number of principles that you've agreed on as the rules of the game. And whenever any statement, any mathematical assertion, involves more information than the amount in those axioms, then it's very natural that it will escape the ability of those axioms.

So you see, the way that mathematics progresses is you trivialize everything! The way it progresses is that you take a result that originally required an immense effort, and you reduce it to a trivial corollary of a more general theory!

Let me give an example involving Fermat's "last theorem", namely the assertion that

$$x^n + y^n = z^n$$

has no solutions in positive integers x, y, z, and n with n greater than 2. Andrew Wiles's recent proof of this is hundreds of pages long, but, probably, a century or two from now there will be a one-page proof! But that one-page proof will require a whole book inventing a theory with concepts that are the natural concepts for thinking about Fermat's last theorem. And when you work with those concepts it'll appear immediately obvious—Wiles's proof will be a trivial afterthought—because you'll have imbedded it in the appropriate theoretical context.

And the same thing is happening with incompleteness.

Gödel's result, like any very fundamental basic result, starts off by being very mysterious and complicated, with a long impenetrable proof. People said about Gödel's original paper the same thing that they said about Einstein's theory of relativity, which is that there are less than five people on this entire planet who understand it. The joke was that Eddington, astronomer royal Sir Arthur Eddington, is at a formal dinner party—this was just after World War I—and he's introduced as one of the three men who understands Einstein's theory. And he says, "Let's see, there's Einstein, and there's me, but who's the other guy?"

So in 1931 Gödel's proof was like that. If you look at his original paper, it's very complicated. The details are programming details we would say now—really it's a kind of complication that we all know how to handle now—but at the time it looked very mysterious. This was a 1931 mathematics paper, and all of a sudden you're doing what amounts to LISP programming, thirty years before LISP was invented! And there weren't even any computers then!

But when you get to Turing, he makes Gödel's result seem much more natural. And I think that my idea of program-size complexity and information—really, algorithmic information content—makes Gödel's result seem more than natural, it makes it seem, I'd say, obvious, inevitable. But of course that's the way it works, that's how we progress.

Where Do We Go from Here?!

I should say, though, that if this were really true, if it were **that** simple, then that would be the end of the field of metamathematics. It would be a sad thing, because it would mean that this whole subject is dead. But I don't think that it is!

You know, I've been giving versions of this talk for many years. In these talks I like to give examples of things that might escape the power of normal mathematical reasoning. And my favorite examples were Fermat's last theorem, the Riemann hypothesis, and the four-color conjecture. When I was a kid these were the three most outstanding open questions in all of mathematics.

But a funny thing happened. First the four-color conjecture was settled by a computer proof, and recently the proof has been greatly improved. The latest version has more ideas and less computation, so that's a big step forward. And then Wiles settled Fermat's last theorem. There was a misstep, but now everyone's convinced that the new proof is correct.

In fact, I was at a meeting in June 1993, when Wiles was presenting his proof in Cambridge. I wasn't there, but I was at a meeting in France, and the word was going around by e-mail that Wiles had done it. It just so happened that I was session chairman, and at one point the organizer of the whole meeting said, "Well, there's this rumor going around, why don't we make an announcement. You're the session chairman, you do it!" So I got up and said, "As some of you may have heard, Andrew Wiles has just demonstrated Fermat's last theorem." And there was **silence!** But afterwards two people came up and said, "You were joking, weren't you?" And I said, "No, I wasn't joking." It wasn't April 1st!

Fortunately the Riemann hypothesis is still open at this point, as far as I know!

But I was using Fermat's last theorem as a possible example of incompleteness, as an example of something that might be beyond the power of the normal mathematical methods. I needed a good example, because people used to say to me, "Well, this is all very well and good, AIT is a nice theory, but give me an example of a specific mathematical result that you think escapes the power of the usual axioms." And I would say, well, maybe Fermat's last theorem!

So there's a problem. Algorithmic information theory is very nice and shows that there are lots of things that you can't prove, but what about individual mathematical questions? How about a natural mathematical question? Can these methods be applied? Well, the answer is no, my methods are not as general as they sound. There are technical limitations. I can't analyze Fermat's last theorem with these methods. Fortunately! Because if I had announced that my methods show that Fermat's last theorem can't be settled, then it's very embarrassing when someone settles it!

So now the question is, how come in spite of these negative results, mathematicians are making so much progress? How come mathematics works so well in spite of incompleteness? You know, I'm not a pessimist, but my results have the wrong kind of feeling about them, they're much too pessimistic!

So I think that a very interesting question now is to look for positive results... There are already too many negative results! If you take them at face value, it would seem that there's no way to do mathematics, that mathematics is impossible. Fortunately for those of us who do mathematics, that doesn't seem to be the case. So I think that now we should look for positive results... The fundamental questions, like the questions of philosophy, they're great, because you never exhaust them. Every generation takes a few steps forward... So I think there's a lot more interesting work to be done in this area.

And here's another very interesting question: Program size is a complexity measure, and we know that it works great in metamathematics, but does it have anything to do with complexity in the real world? For example, what about the complexity of biological organisms? What about a theory of evolution?

Von Neumann talked about a general theory of the evolution of life. He said that the first step was to define complexity. Well, here's a definition of complexity, but it doesn't seem to be the correct one to use in theoretical biology. And there is no such thing as theoretical biology, not yet!

As a mathematician, I would love it if somebody would prove a general result saying that under very general circumstances life has to evolve. But I don't know how you define life in a general mathematical setting. We know it when we see it, right? If you crash into something alive with your car, you know it! But as a mathematician I don't know how to tell the difference between a beautiful deer running across the road and the pile of garbage that my neighbor left out in the street! Well, actually that garbage is connected with life, it's the debris produced by life...

So let's compare a deer with a rock instead. Well, the rock is harder, but that doesn't seem to go to the essential difference that the deer is alive and the rock is a pretty passive object. It's certainly very easy for us to tell the difference in practice, but what is the fundamental difference? Can one grasp that mathematically?

So what von Neumann was asking for was a general mathematical theory. Von Neumann used to like to invent new mathematical theories. He'd invent one before breakfast every day: the theory of games, the theory of self-reproducing automata, the Hilbert space formulation of quantum mechanics... Von Neumann wrote a book on quantum mechanics using Hilbert spaces—that was done by von Neumann, who had studied under Hilbert, and who said that this was the right mathematical framework for doing quantum mechanics.

Von Neumann was always inventing new fields of mathematics, and since he was a childhood hero of mine, and since he talked about Gödel and Turing, well, I said to myself, if von Neumann could do it, I think I'll give it a try. Von Neumann even suggested that there should be a theory of the complexity of computations. He never took any steps in that direction, but I think that you can find someplace where he said that this has got to be an interesting new area to develop, and he was certainly right.

Von Neumann also said that we ought to have a general mathematical theory of the evolution of life... But we want it to be a very general theory, we don't want to get involved in low-level questions like biochemistry or geology... He insisted that we should do things in a more general way, because von Neumann believed, and I guess I do too, that if Darwin is right, then it's probably a very general thing.

For example, there is the idea of *genetic programming,* that's a computer version of this. Instead of writing a program to do something, you sort of evolve it by trial and error. And it seems to work remarkably well, but can you prove that this has got to be the case? Or take a look at Tom Ray's *Tierra...* Some of these computer models of biology almost seem to work too well—the problem is that there's no theoretical understanding why they work so well. If you run Ray's model on the computer you get these parasites and hyperparasites, you get a whole ecology. That's just terrific, but as a pure mathematician I'm looking for theoretical understanding, I'm looking for a general theory that starts by defining what an organism is and how you measure its complexity, and that **proves** that organisms have to evolve and increase in complexity. That's what I want, wouldn't that be nice?

And if you could do that, it might shed some light on how general the phenomenon of evolution is, and whether there's likely to be life elsewhere in the universe. Of course, even if mathematicians never come up with such a theory, we'll probably find out by visiting other places and seeing if there's life there... But anyway, von Neumann had proposed this as an interesting question, and at one point in my deluded youth I thought that maybe program-size complexity had something to do with evolution... But I don't think so anymore, because I was never able to get anywhere with this idea...

So I think that there's a lot of interesting work to be done! And I think that we live in exciting times. In fact, sometimes I think that maybe they're even a little bit too exciting!... And I hope that if this talk were being given a century from now, in 2099, there would be **another** century of exciting controversy about the foundations of mathematics to summarize, one with different concerns and preoccupations... It would be interesting to hear what that talk would be like a hundred years from now! Maybe some of you will be there! Or give the talk even! Thank you very much!

Further Reading

1. G. J. Chaitin, *The Unknowable,* Springer-Verlag, 1999.

2. G. J. Chaitin, *The Limits of Mathematics,* Springer-Verlag, 1998.

A century of controversy over the foundations of mathematics

G. J. Chaitin's 2 March 2000 Carnegie Mellon University School of Computer Science Distinguished Lecture. The lecture was videotaped; this is an edited transcript.

We're in a state of euphoria now in the computer business because things are going so well: the web, e-commerce. It's all paying for our salaries, and it's a nice moment to be around, when things are going so well. But I'd like to make the outrageous claim, that has a little bit of truth, that actually all of this that's happening now with the computer taking over the world, the digitalization of our society, of information in human society, you could say in a way is the result of a philosophical question that was raised by David Hilbert at the beginning of the century.

It's not a complete lie to say that Turing invented the computer in order to shed light on a philosophical question about the foundations of mathematics that was asked by Hilbert. And in a funny way that led to the creation of the computer business.

It's not completely true, but there is some truth in it. You know, most historical statements are a lie, so this one isn't that much worse than most others!

So I'd like to explain the philosophical history of the computer. In a way what happened, and I'll tell you more, is that Hilbert said we should formalize all of mathematics, mathematical reasoning. And this failed: it took Gödel and Turing to show that it couldn't be done. It failed in that precise technical sense. But in fact it succeeded magnificently, not formalization of reasoning,

129

but formalization of algorithms has been the great technological success of our time — computer programming languages!

So if you look back at the history of the beginning of this century you'll see papers by logicians studying the foundations of mathematics in which they had programming languages. Now you look back and you say this is clearly a programming language! If you look at Turing's paper of course there's a machine language. If you look at papers by Alonzo Church you see the lambda calculus, which is a functional programming language. If you look at Gödel's original paper you see what to me looks like LISP, it's very close to LISP, the paper begs to be rewritten in LISP!

So I'd like to give you this hidden philosophical history of computer technology which is how philosophically minded mathematicians set out to solve once and for all the foundational problems of mathematics and did not succeed but helped to create computer technology as a by product. This was the failure of this project! We're all benefiting from the glorious failure of this project!

However this project has not died completely. — I'm going to start more systematically from the beginning; but I'm trying to give an introduction. — It's popular to think, well Gödel did this wonderful thing in 1931 and Turing added a lot of profound stuff in 1936, but the world has moved on from that point. And what I'd like to do is to tell you that in fact I've done some more work in this area.

You may think it's misguided! Most of the world has shrugged and gone on. We had this disappointment. What Gödel and Turing showed is that axiomatic formal reasoning has certain limitations. You can't formalize it all. And at first people were tremendously shocked and then they shrugged and said, so what? Mathematicians went on, ignoring this. And my misfortune or fortune was that I didn't want to shrug. I said, I want to understand this better. And I'm going to tell you the story of my attempt to understand Gödel incompleteness. — It's a psychological problem that a good psychiatrist could have cured me of, and then I wouldn't have done any of this work!

So let me start at the beginning and tell you this story of a hundred years of intense worry, crisis, self-doubt, self-examination and angst about the philosophy of mathematics.

There've been lots of crises in the history of mathematics. Mathematics is not placid, static and eternal.

One of the first crises was the Pythagorean result that the square root of two is **irrational**. And the fact that this was a crisis survives in the word "irrational". Remember the Greeks thought that rationality was the supreme goal — Plato! Reason! If a number is called irrational that means that this was the Gödel incompleteness theorem of ancient Greece. So there was a crisis there.

Another crisis was caused by the calculus. A lot of people said this is nonsense, we're talking about infinitesimals, what is this? Bishop Berkeley was a theologian and he said, pure mathematicians make as little sense as theologians, you can't reject us saying **we're** unreasonable. The way you deal with evanescent quantities in the calculus — this was before the calculus had a rigorous foundation — is as bad as our theological discussions! So at that time it was pretty bad!

Then there was a crisis about the parallel postulate, about non-Euclidean geometries.

So mathematics is not static and eternal!

But the particular crisis that I want to tell you about goes back a little more than a hundred years to work of Cantor on set theory.

Cantor: Theory of **Infinite** Sets

So my talk is very impractical. We all know that you can have a start-up and in one year make a million dollars if you're lucky with the web. So this is about how not to make any money with the web. This is about how to ruin your career by thinking about philosophy instead.

So Cantor was obsessed with the notion of infinite, and it's not mentioned that he was obsessed with infinite because he was interested in theology and God, which is edited out from the accounts now, but that was the original idea.

And Cantor had the idea that if you have 1, 2, 3,... why stop there?

$$1, 2, 3, \ldots \omega$$

— I'm giving you a cartoon version of Cantor's theory of infinite sets. — You put an omega, ω, this is a Greek letter, the lower case of the last letter in the Greek alphabet, that's the reason to pick it. So you just say, I'm going to put another number here instead of stopping with 1, 2, 3,... This is going to be the first number after all the finite numbers. This is the first transfinite number.

You can keep going for a while.

$$1, 2, 3, \ldots \omega, \omega + 1, \omega + 2, \ldots$$

And then you have another thing like a copy of 1, 2, 3,... : $\omega + 1$, $\omega + 2$, $\omega + 3$, ... These are names. And then you say, why stop here? I'm going to put something after all this, so 2ω, $2\omega + 1$, $+2$, $+3$, then later 3ω, 4ω... Well, what comes after all of those? Why stop there? So, ω squared, obviously.

$$1, 2, 3, \ldots \omega, \omega + 1, \omega + 2, \ldots 2\omega \; 3\omega \; 4\omega \; \omega^2$$

Then you keep going. $5\omega^2 + 8\omega + 96$! And then much later you get to ω cubed! And then eventually ω to the fourth. You keep going and why stop there? This sequence goes on forever, but let's put something after all of those. So what would that be? That would be obviously ω to the ω. This is starting to get interesting! Then you keep going and you have ω to the ω to the ω. This is a pretty far-out number already!

$$1, 2, 3, \ldots \omega, \omega + 1, \omega + 2, \ldots 2\omega \; 3\omega \; 4\omega \; \omega^2 \; \omega^3 \; \omega^4 \; \omega^\omega \; \omega^{\omega^\omega}$$

You can see why this is becoming theological. This is the mathematical equivalent of drug addiction. Instead of getting high on alcohol or grass you get high on ideas like this. After a while you don't know where you're standing or what's going on!

Then the next number is ω to the ω to the ω forever.

$$\omega^{\omega^{\omega^{\omega^{\omega^{\omega^{\cdots}}}}}}$$

This number is the smallest solution of the equation

$$x = \omega^x$$

And it's called ϵ_0, epsilon nought, I don't know why. Because you start having problems with how to name things, because up to here I was using normal algebraic notation just throwing in ω.

So anyway you can see this is fantastic stuff! I don't know whether it's mathematics, but it's very imaginative, it's very pretty, and actually there was a lot of practical spin-off for pure mathematicians from what Cantor was doing.

Some people regarded set theory as a disease. Poincaré, the great French mathematician, said set theory is a disease, he said, from which I hope future generations will recover. But other people redid all of mathematics using the set-theoretic approach. So modern topology and a lot of abstract mathematics of the twentieth century is a result of this more abstract set-theoretic approach, which generalized questions. The mathematics of the nineteenth century was at a lower level in some ways, more involved with special cases and formulas. The mathematics of the twentieth century — it's hard to write a history of mathematics from the year ten-thousand looking back because we're right here — but the mathematics of the twentieth century you could almost say is set-theoretical, "structural" would be a way to describe it. The mathematics of the nineteenth century was concerned with formulas, infinite Taylor series perhaps. But the mathematics of the twentieth century went on to a set-theoretic level of abstraction.

And in part that's due to Cantor, and some people hate it saying that Cantor wrecked and ruined mathematics by taking it from being concrete and making it wishy-washy, for example, from hard analysis to abstract analysis. Other people loved this. It was very controversial.

It was very controversial, and what didn't help is in fact that there were some contradictions. It became more than just a matter of opinion. There were some cases in which you got into really bad trouble, you got obvious nonsense out. And the place you get obvious nonsense out in fact is a theorem of Cantor's that says that for any infinite set there's a larger infinite set which is the set of all its subsets, which sounds pretty reasonable. This is Cantor's diagonal argument — I don't have time to give you the details.

So then the problem is that if you believe that for any infinite set there's a set that's even larger, what happens if you apply this to the universal set, the set of everything? The problem is that by definition the set of everything has everything, and this method supposedly would give you a larger set, which is the set of all subsets of everything. So there's got to be a problem, and the problem was noticed by Bertrand Russell.

Bertrand Russell

Cantor I think may have noticed it, but Bertrand Russell went around telling everyone about it, giving the bad news to everyone! — At least Gödel attributes to Russell the recognition that there was a serious crisis.

The disaster that Russell noticed in this proof of Cantor's was the set of all sets that are not members of themselves, that turns out to be the key step in the proof. And the set of all sets that aren't members of themselves sounds like a reasonable way to define a set, but if you ask if it's inside itself or not, whatever you assume you get the opposite, it's a contradiction, it's like saying this statement is false. The set of all sets that are not members of themselves is contained in itself if and only if it's not contained in itself.

So does this mean that some ways of defining sets are bad, or that the universal set gets you into trouble? What's wrong with the set of everything? So there was a problem with set theory — that became increasingly clear. I think Russell helped to make it be recognized by everybody that we had a serious crisis and that methods of reasoning that seemed at first sight perfectly legitimate in some cases led to obvious disaster, to contradictions. There were a whole bunch of paradoxes that Russell advertised: the Berry paradox, the one I just mentioned is called the Russell paradox, and there's another paradox, the Burali-Forti paradox.

A lot of these paradoxes in fact were really brought to the attention of the world by Russell. Russell would typically have a footnote saying this paradox occurred to me while I was reading a paper by Burali-Forti, so everyone calls it the Burali-Forti paradox. Burali-Forti I think spent his whole life trying to live down this attribution because he didn't believe that mathematics was in trouble!

Okay so there was a crisis, and I think Russell was one of the key figures in this. At this point David Hilbert comes to the rescue.

David Hilbert

David Hilbert was a very important mathematician around the turn of the century. Unlike Poincaré, a very important French mathematician — Hilbert was a very important German mathematician — Hilbert liked set theory. He liked this abstract Cantorian approach. And Hilbert had the idea of solving once and for all these problems. How was he going to do it?

The way Hilbert was going to do it is using the axiomatic method, which of course goes back to Euclid — Hilbert didn't invent this. But he went one significant step further.

Hilbert: **Formal** Axiomatic Method

Hilbert said let's use all the technology from symbolic logic, which a lot of people were involved in inventing, and let's go to some final extreme. Because one of the reasons you got into trouble and got contradictions in mathematics with set theory is because words are very vague. What we want to do to get rid of all these problems in mathematics and in reasoning is get rid of pronouns for example, you don't know what pronouns refer to. And there are all kinds of things that are vague in normal language.

Hilbert said that the way to get rid of all these problems is to come up with a finite set of axioms and an artificial language for doing mathematics — this is the idea of formalism taken to the limit.

Formalism

Take formalism to the absolute limit and invent a completely artificial language with completely precise rules of the game — artificial grammar and everything — and eliminate all these problems, like the problems that Russell had. This was an ambitious program to once and for all put mathematics on a firm footing.

And one thing that Hilbert emphasized, which was as far as I know a key contribution that he himself made, was that he wanted the rules of the game for this formal axiomatic system for all of mathematics to be so precise that you have a mechanical proof checker. So it's completely certain and objective and mechanical whether a proof obeys the rules or not. There should be no human element, there should be no subjective element, there should be no question of interpretation. If somebody claims they have a proof, it should be absolutely clear, mechanical, to check it and see, does it obey the rules and you proved a theorem or does it have a mistake, does it fail.

So this is the idea that mathematics should be absolutely black or white, precise, absolute truth. This is the traditional notion of mathematics.

Black or White

The real world we know is an absolute mess — right? — everything's complicated and messy. But the one place where things should be absolutely clear, black or white, is in pure mathematics.

So this is sort of what Hilbert is saying, and he proposed this as a goal, to have this formalization of all of mathematics and eliminate all the problems. Now this was a program, this was not supposed to be something you did over a weekend. Hilbert proposed this as a goal for putting mathematics on a very firm foundation. And he and a group of very bright collaborators, including John von Neumann, set to work on this, and for a while, for thirty years, it looked sort of encouraging. And then — this is a quick summary of a century of work — then as I'm sure all of you know there were a few little problems!

The problems are 1931, Kurt Gödel, and 1936, Alan Turing.

1931 Gödel

1936 Turing

They showed that it could not be done, that there were fundamental obstacles to formalizing all of mathematics and making mathematics absolutely black and white and absolutely crystal clear. Remember what Hilbert is proposing is that we should formalize all of mathematics so that everyone on planet earth can agree that a proof is either correct or incorrect. The rules of the game should be absolutely explicit, it should be an artificial language and then mathematics will give you **absolute truth**. "Absolute truth" should be underlined in a very beautiful font and you should hear the angels singing when you say these words! This was the thought that we mathematicians have absolute truth. It's ours — no one else has it, only us! That was the idea.

So it turns out this doesn't quite work. Why doesn't it work?

Gödel shocked people quite a bit by showing that it couldn't work. It was very, very surprising when Gödel did this in 1931. And Turing went I think more deeply into it. So let me give you a cartoon five minute summary, my take on what they did.

Gödel starts with "this statement is false", what I'm now saying is a lie, I'm lying. If I'm lying, and it's a lie that I'm lying, then I'm telling the truth! So "this statement is false" is false if and only if it's true, so there's a problem. Gödel considered instead "this statement is unprovable".

"This statement is unprovable!"

Here unprovable means unprovable from the axioms of Hilbert's formal axiomatic system, unprovable within the system that Hilbert was trying to create.

Now think about a statement that says that it's unprovable. There are two possibilities: it's provable or it's unprovable. This is assuming you can make a statement say it's unprovable, that there's some way to say this within Hilbert's system. That required enormous cleverness: Gödel numbering, trickery for a statement to refer to itself indirectly, because pronouns that say "this" or "I" aren't usually found in mathematical formulas. So this required a lot of cleverness on Gödel's part. But the basic idea is "this statement is unprovable".

So there are two possibilities. Either it's provable or it's unprovable. And this means provable or unprovable from the system that Hilbert had proposed, the final goal of formalizing all of mathematics.

Well, if it's provable, and it says it's unprovable, we're proving something that's false. So that's not very nice. And if it's unprovable and it says it's unprovable, well then, what it states is true, it's unprovable, and we have a hole. Instead of proving something false we have incompleteness, we have a true statement that our formalization has not succeeded in capturing.

So the idea is that either we're proving false statements, which is terrifying, or we get something which is not as bad, but is still awful, which is that our formal axiomatic system is incomplete — there's something that's true but we can't prove it within our system. And therefore the goal of formalizing once and for all all of mathematics ends up on the floor!

Now I don't think that Hilbert really wanted us to formalize all of mathematics. He didn't say that we should all work in an artificial language and have formal proofs. Formal proofs tend to be very long and inhuman and hard to read. I think Hilbert's goal was philosophical. If you believe that mathematics gives absolute truth, then it seems to me that Hilbert has got to be right, that there ought to have been a way to formalize once and for all all of mathematics. That's sort of what mathematical logic was trying to do, that's sort of what the axiomatic method was trying to do, the idea of breaking proofs into smaller and smaller steps. And Leibniz thought about this, and Boole thought about this, and Frege and Peano and Russell and Whitehead thought about this. It's the idea of making very clear how mathematics operates step by step. So that doesn't sound bad. Unfortunately it crashes at this point!

So everyone is in a terrible state of shock at this point. You read essays by Hermann Weyl or John von Neumann saying things like this: I became a mathematician because this was my religion, I believed in absolute truth, here was beauty, the real world was awful, but I took refuge in number theory. And all of a sudden Gödel comes and ruins everything, and I want to kill myself!

So this was pretty awful. However, this

"This statement is unprovable!"

is a very strange looking statement. And there are ways of rationalizing, human beings are good at that, you don't want to face unpleasant reality. And this unpleasant reality is very easy to shrug off: you just say, well, who cares! The statements I work with normally in mathematics, they're not statements of this kind. This is nonsense! If you do this kind of stupidity, obviously you're going to get into trouble.

But that's rationalizing too far. Because in fact Gödel made this

"This statement is unprovable!"

into a statement in elementary number theory. In its original form, sure, it's nonsense, who ever heard of a statement in mathematics that says it's unprovable? But in fact Gödel made this into a numerical statement in elementary number theory, in arithmetic. It was a large statement, but in some clever way, involving Gödel numbering of all arithmetic statements using prime numbers, he was writing it so that it looked like a statement in real mathematics. But it really indirectly was referring to itself and saying that it's unprovable.

So that's why there's a problem. But people didn't really know what to make of this. So I would put "surprising" here, surprising, a terrible shock!

1931 Gödel "This statement is unprovable!" **Surprising**

Now my reaction as a child reading this proof is that I follow it step by step, but I don't like it. It doesn't appeal to me! Which is good, because if I had said I like it, it's wonderful, finished, I go ahead and become a molecular biologist and start up a biotech company, and now I'd be rich, but I wouldn't have done any work in this area!

Then comes Turing.

1936 Turing

Now I prefer Turing's approach. Turing goes more deeply into this. Turing starts talking about computers. This is the point where it happens!

1936 Turing **Computer**

Turing has to invent the computer, because Hilbert says that there should be a mechanical procedure to decide if a proof is correct or not. Turing says what Hilbert really means is that there should be a computer program for checking proofs. But first Turing has to say what a computer is, it's a Turing machine, and all of this is in a paper of Turing's in 1936, when there were no computers, so it's a fantastic piece of work. And I would like to claim that this is the invention of the computer. These were general-purpose computers, that was the idea, on paper.

What Turing shows is in fact that there is a relatively concrete statement that escapes the power of mathematics. We now think of computers as physical devices, so they're almost like something in physics. It's a machine working away, it's an idealization of that, you have this machine working, and Turing discovers the halting problem.

1936 Turing **Computer** Halting problem

The halting problem says there's no way to decide if a computer program will **eventually** halt.

Now obviously to decide if a computer program halts is the easiest thing in the world. You run it and when you run out of patience, that's it, it doesn't halt as far as you're concerned. Who cares, you can't wait any longer! But what Turing showed is that there's a problem if you put no time limit. This is very abstract mathematics — in the real world there's always a time limit! You can't run a program a million years, a billion years, $10^{10^{10}}$ years! If you put a time limit, the halting problem is very easy to decide, in principle: you just run the program that long and you see, does it halt by that point or not.

But what Turing showed is that if you put no time limit, then there is no solution. There's no way to decide in advance whether a computer program will halt or not. If it halts you can eventually discover that by running it. The problem is to realize that you've got to give up. So there's no mechanical procedure that will decide in advance if a computer program will halt or not, and therefore, it turns out, there is no set of mathematical axioms in Hilbert's sense that can enable you to prove whether a program will halt or not.

Because if you could always prove whether a program will halt or not, you could run through all possible proofs in size order and check whether they're correct, and eventually either find a proof that the program's going to halt or find a proof that it's not going to halt. And this would give you a way to decide in advance whether a program's going to halt.

Now in practice running through all possible proofs requires an astronomical amount of time. Imagine how many proofs are there that are one page long! You'd never get through them! But in principle you can run through all possible proofs in size order and check whether they obey the rules, if it's a Hilbert formal axiomatic system. So if you had a formal axiomatization of mathematics that enabled you to always prove whether a program halts or not, that would give you a mechanical procedure, by running through all possible proofs in size order, to decide whether a program will halt or not. And Turing showed that you can't do it. His proof, by the way, involves Cantor's diagonal argument — all these ideas are connected, but there's no time to go into that.

So I think that Turing's work makes the limits of mathematics seem much more natural, because we're talking about a question about a physical device, it's a computer.

1936 Turing **Computer** Halting problem **Natural**

You fantasize a little bit, you make it a theoretical computer, a computer that can go on forever, that never breaks down, that has as much storage as it wants, so that if numbers get too big it can keep going anyway. But that's not too much of a fantasy; we have devices like that everywhere, right? So it sounds much more concrete. The limits of mathematics discovered by Turing sound more serious, more dangerous than the ones that Gödel found.

And this is the invention of the computer, for this crazy kind of theoretical argument! You don't see billions and billions of dollars of technology in this 1936 paper, but it was all there in embryonic form, as von Neumann kept emphasizing: the universal Turing machine is really the notion of a general-purpose programmable computer. You had machines that did calculations before, but they did special-purpose calculations, they were adding machines, mechanical calculating machines, and I used them when I was a kid. But the notion of a computer is Turing's notion of a machine that can do what any calculating machine can do, and that's the idea of software: it's a very general-purpose machine, it's a flexible machine. So it's really there, von Neumann kept saying, very clearly in Turing's paper. So you have this whole technology there!

And in fact Gödel's paper as I said uses LISP, there's a programming language hidden in it, and in Turing's paper there's a programming language, given explicitly, Turing machines, and it's a machine language. It's actually a very bad machine language, it's a machine that no person in their sane mind would want to program. But Turing wanted to keep it as simple as possible. Obviously, if his paper had included a manual for the machine language of a real machine, it would have been hopeless, no one would have understood it.

Okay, now what happens with all of this? What happens with all of this is that Hilbert dies, World War II comes, and when I'm a child in the 1950's I could still read essays by John von Neumann talking about all of this, but the world was clearly going in a less philosophical direction. Things were going downhill rapidly until we're all billionaires with our web start-ups! People were less concerned about philosophy, and computers were becoming a technology, and Turing was very involved in that, and so was von Neumann.

But stupidly I wanted to understand what was going on in the foundations of mathematics, so in a way I'm stuck in the 1930's, I never got past that stage. What happened? What happened with me is that I couldn't accept the fact that everybody said, who cares! Now it's true that there are a lot of things in life besides the foundations of mathematics and epistemology! There're things like having a family, earning a living, wars, politics, lots of stuff out there, obviously! But what I couldn't accept was that even in the world of pure mathematics, mathematicians were saying, so what, in practice we should do mathematics exactly the same as we've always done it, this does not apply to the problems I care about! That was basically the reaction to Gödel's and Turing's work on incompleteness.

At first there was terrible shock, then it went from one extreme to another. Who cares, people would say, it's obvious, or it's irrelevant! This has no impact in practice on how we should do mathematics. I was very unhappy with that. I was obsessed by incompleteness, and I had an idea.

When I was a kid I really wanted to be a physicist, and a lot of mathematicians say I never made it into mathematics really — I never succeeded, I'm still stuck! I wanted to be a physicist, and I got corrupted by a lot of ideas from physics. While all of this crisis was going on in mathematics, there was a parallel crisis going on in physics, which actually started in the 1920's: that's quantum mechanics, and the key date is 1924.

1924 Quantum Mechanics

And that's the whole question of uncertainty and randomness in fundamental physics. So when I was a kid, besides reading essays talking about Gödel's incompleteness theorem saying "Oh, my God", there were also essays asking what happened to determinism in physics, what happened to predictability, can there be randomness, does God play dice? Einstein said no, God doesn't play dice. He hated quantum mechanics. And everybody else said yes, God plays dice.

God plays dice!

Quantum mechanics is the most successful physical theory ever. We get transistors and computers from it. But even though Einstein helped to contribute to the creation of quantum mechanics he hated it. So it looks like Einstein was wrong. God does play dice!

So I had a crazy idea. I thought that maybe the problem is larger and Gödel and Turing were just the tip of the iceberg. Maybe things are much worse and what we really have here in pure mathematics is randomness. In other words, maybe sometimes the reason you can't prove something is not because you're stupid or you haven't worked on it long enough, the reason you can't prove something is because there's nothing there! Sometimes the reason you can't solve a mathematical problem isn't because you're not smart enough, or you're not determined enough, it's because there is no solution because maybe the mathematical question has no structure, maybe the answer has no pattern, maybe there is no order or structure that you can try to understand in the world of pure mathematics. Maybe sometimes the reason that you don't see a pattern or structure is because there is no pattern or structure!

And one of my motivations was the prime numbers. There's some work on the prime numbers that says that in some ways the prime numbers can be looked at statistically. There seems to be a certain amount of randomness in the distribution of the primes. That's one of the ways that people try to think about the prime numbers. And this even happens in number theory, which is the queen of pure mathematics!

So on the one hand I heard this talk about probabilistic ways of thinking about the primes — this was heuristic — and this stuff about God plays dice in fundamental physics — what goes on in the atom is random — and I begin to think, well, maybe that's what's going on in the foundations of mathematics.

This is what I set out to do, and this project took a long time. One of the first steps is clarifying what do you mean by randomness. What do you mean by lack of structure, lack of order, lack of pattern?

Randomness: Lack of structure

So this is a kind of a logical notion of randomness rather than a statistical notion of randomness. It's not like in physics where you say a physical process is random like coin tossing. I don't care where something comes from. I just look at something and say does it have structure or pattern or not. So this is logical or structural randomness as opposed to physical unpredictability and randomness. It's different — it's very closely related, but they're different.

And the idea that I came up with — and Kolmogorov came up with at the same time independently — is the idea that something is random if it can't be compressed into a shorter description, if essentially you just have to write it out as it is. In other words, there's no concise theory that produces it. For example, a set of physical data would be random if the only way to publish it is as is in a table, but if there's a theory you're compressing a lot of observations into a small number of physical principles or laws. And the more the compression, the better the theory: in accord with Occam's razor, the best theory is the simplest theory. I would say that a theory is a program — also Ray Solomonoff did some thinking along these lines for doing induction — he didn't go on to define randomness, but he should have! If you think of a theory as a program that calculates the observations, the smaller the program is relative to the output, which is the observations, the better the theory is.

By the way, this is also what axioms do. I would say that axioms are the same idea. You have a lot of theorems or mathematical truth and you're compressing them into a set of axioms. Now why is this good? Because then there's less risk. Because the axioms are hypotheses that you have to make and every time you make a hypothesis you have to take it on faith and there's risk — you're not proving it from anything, you're taking it as a given, and the less you assume, the safer it is. So the fewer axioms you have, the better off you are. So the more compression of a lot of theorems, of a body of theory, into a small set of axioms, the better off you are, I would say, in mathematics as well as physics.

Okay, so this is this notion of lack of structure or randomness. You have to define it first! If I'm going to find randomness or lack of structure, lack of pattern, in pure mathematics, first I've got to say what do I mean by that. And I like to call this subject algorithmic information theory. It deals with this algorithmic information. Or you can call it complexity if you like, program-size complexity.

Algorithmic Information

The basic concept is to look at the size of the most concise program, the smallest program — I don't care about running time — it's the most concise program that calculates something. That's the number of bits I have to give a computer in order to get it to produce this object. That's my most concise algorithmic description of something, and that's how I measure its complexity, its algorithmic information content or its program-size complexity.

This is like recursive function theory: I don't care about run time — so this is very impractical! So in that sense also what I'm doing is 1930's stuff, with this one extra idea thrown in of program size, of looking at the size of programs.

So what happens when you start looking at the size of programs? — and then something is random if the smallest program that calculates it is the same size as it is, and there's no compression. So the whole idea is, look at the size of computer programs, don't care about run time — if it takes a billion, billion years I don't care! Information is the only thing I'm thinking about, bits of information, size of computer programs. Okay?

So what happens when you start playing with this idea? What happens is, everywhere you turn, you get incompleteness and undecidability, and you get it in the worst possible way. For example this happens with the first thing you want to do: you can never decide that an individual string of digits satisfies this definition of randomness. Impossible! You can never calculate the program-size complexity of anything. You can never determine what the size of the smallest program is.

If you have a program that calculates something, that gives you an **upper** bound, its size is an upper bound on the program-size complexity of what it calculates. But you can never prove any **lower** bounds. And that's my first incompleteness result in this area and I think Jack Schwartz got very excited about it.

In normal, practical, useful complexity theory where you talk about time rather than bits of information, lower bounds are much harder than upper bounds. To get lower bounds on complexity is much harder than getting upper bounds on complexity. Because if you find a clever algorithm you get an upper bound on the time it takes to calculate something; if you find a way to do it that's fast you've shown that it can be done that fast. The problem is to show that you've gotten the fastest possible algorithm, that's much harder, right? But it can be done in some cases, within a class of possible algorithms. Well, in algorithmic information theory you can't prove **any** lower bounds! And I had an article about this in 1975 in *Scientific American*.

The basic idea is that you can't prove any lower bounds on the program-size complexity of individual objects. So in particular even though most strings of digits satisfy this definition of randomness, they're incompressible in this sense, they're random in this sense of lack of structure — it turns out you can show easily that most objects satisfy this definition, they have no structure — if you look at all hundred digit numbers, almost all of them have no structure according to this definition, but you can never be sure in individual cases, you can never prove it in individual cases.

More precisely, there may be finitely many exceptions. With N bits of axioms you can determine all the objects of program-size complexity up to N. But that's as far as you can go.

And my worst incompleteness result, my very worst incompleteness result, where you have complete lack of structure in pure mathematics, has to do with a number I defined called the halting probability.

$$\Omega = \text{halting probability}$$

How is this number defined? It's very simple. Turing said you can't decide whether a program halts, there's no mechanical procedure for doing that. And I say, let's consider a real number Ω which is the probability that a program generated by tossing a coin halts. So I'm averaging over Turing's halting problem, saying if I generate a program by coin tossing, what is the probability that it halts, with no time limit? So this will give me a real number that's determined if you tell me — there's a subscript — what's the programming language.

$$\Omega_{\text{computer}} = \text{halting probability of computer}$$

Once you decide, then Ω is a well-defined real number. Mathematically it's not a very sophisticated thing! Compared to large cardinals, sophisticated mathematics, this is a fairly low-brow object.

However it turns out this object Ω is maximally unknowable!

Ω is **maximally unknowable**

What is it that's maximally unknowable? Well, it's the digits or bits of this number. Once I fix the computer programming language this halting probability is a specific real number, that depends on the choice of computer, or the programming language in which I generate a program by coin tossing. So this becomes a specific real number, and let's say I write it out in binary, so I get a sequence of 0's and 1's, it's a very simple-minded definition. Well, it turns out these 0's and 1's have **no** mathematical structure. They cannot be compressed. To calculate the first N bits of this number in binary requires an N-bit program. To be able to prove what the first N bits of this number are requires N bits of axioms. This is irreducible mathematical information, that's the key idea.

Ω is **irreducible information**

This should be a shocking idea, irreducible mathematical information, because the whole normal idea of mathematics, the Hilbertian idea, the classical idea of mathematics, is that all of mathematical truth can be reduced to a small set of axioms that we can all agree on, that are "self-evident" hopefully. But if you want to determine what the bits of the halting probability Ω are, this is something that cannot be reduced to anything simpler than it is.

Ω has a mathematical definition with a rather simple structure once I specify the computer, or the programming language, I've even written out a program in LISP that calculates this number in a weak sense. You can't calculate this number. If you could calculate it, then it wouldn't be unknowable! You can get it in the limit from below, but it converges very, very slowly — you can never know how close you are — there is no computable regulator of convergence, there is no way to decide how far out to go to get the first N bits of Ω right. To get Ω in the limit from below, you just look at more and more programs, for more and more time, and every time you see that a K-bit program halts, that contributes $1/2^K$ to the halting probability.

$$\Omega = \sum_{p \text{ halts}} 2^{-|p|}$$

So the time you need to get the first N bits of Ω right grows like the longest possible finite run-time of an N-bit program, which is a version of the Busy-Beaver function.

So what's the precise definition of Ω? Generate a program by tossing a coin for each bit, that's independent tosses of a fair coin. The key point is that the program has to be "self-delimiting". The computer has got to ask for each bit one by one. Every time the computer says I want another bit of the program, you flip the coin. And the computer has to decide by itself that it has enough bits, that it has the whole program. The program has to be self-delimiting to define this probability measure correctly. So there's no blank to indicate where a program ends: a program has to indicate within itself how long it is with some trick, some coding trick. That's the technical issue to get this probability to be well-defined. That's the one technical point in my theory.

So this number Ω is a real number between 0 and 1. It's the probability that a program each of whose bits is generated by an independent toss of a fair coin eventually halts. And I'm fixing the programming language, I pick the universal Turing machine, there's a subscript, it's Ω_{UTM}, it's the halting probability of a particular universal Turing machine. And I actually pick a particular UTM that I programmed in LISP, just to fix the ideas. But you could do it with essentially any universal Turing machine with self-delimiting programs; it would work.

So Ω is maximally unknowable. This is a case where mathematical truth has no structure or pattern and it's something we're never going to know! So let me tell you what I've got here. What I've got here is maximum randomness — like independent tosses of a fair coin — in pure mathematics. In fact, I can even do it in elementary number theory, like Gödel did. I can make determining bits of Ω into an assertion about a diophantine equation.

The point is, here you've got a simple mathematical question — which is what is each bit of Ω: is the first bit 0 or 1, is the second bit 0 or 1, is the third bit 0 or 1 — but the answers have no structure, they look like independent tosses of a fair coin, even though each answer is well-defined mathematically, because it's a specific bit of a specific real number and it has to be a 0 or a 1. In fact, we're never going to know: this is my version of independent tosses of a fair coin in pure mathematics. Even if you knew all the even bits of Ω it wouldn't help you to get any of the odd bits. Even if you knew the first million bits, it wouldn't help you to get the next one. It really looks like independent tosses of a fair coin, it's maximally random, it has maximum entropy.

Physicists feel comfortable with randomness, but this is the black or white world of pure mathematics — how is this possible, how can it be? Each of these bits is well-defined, it's a specific 0 or a 1, because Ω is a specific real number once I fix the universal Turing machine or the programming language that I'm dealing with. But it turns out that the right way to think about each bit is that it's not black or white, it's not that it's a 0 or a 1, it's so well balanced, it's so delicately balanced, that it's **grey**!

Here's another way to put it. Let's go back to Leibniz. What's the idea of mathematics? The normal idea is that if something is true, it's true for a reason — Leibniz! — if something is true it's true for a reason. Now in pure math, the reason that something is true is called a proof, and the job of the mathematician is to find proofs, to find the reason something is true. But the bits of this number Ω, whether they're 0 or 1, are mathematical truths that are true for **no reason**, they're true **by accident**! And that's why we will never know what these bits are.

In other words, it's not just that Hilbert was a little bit wrong. It's not just that the normal notion of pure mathematics is a little bit wrong, that there are a few small holes, that there are a few degenerate cases like "This statement is unprovable". It's not that way! It's much, much worse than that! There are extreme cases where mathematical truth has no structure at all, where it's maximally unknowable, where it's completely accidental, where you have mathematical truths that are like coin tosses, they're true by accident, they're true for no reason. That's why you can never prove whether individual bits of Ω are 0 or are 1, because **there is no reason** that individual bits are 0 or 1! That's why you can't find a proof. In other words, it's so delicately balanced whether each bit is 0 or 1 that we're never going to know.

So it turned out that not only Hilbert was wrong, as Gödel and Turing showed... I want to summarize all of this. With Gödel it looks surprising that you have incompleteness, that no finite set of axioms can contain all of mathematical truth. With Turing incompleteness seems much more natural. But with my approach, when you look at program size, I would say that it looks inevitable. Wherever you turn, you smash up against a stone wall and incompleteness hits you in the face!

> Program-size complexity & Ω & irreducible information
> $\rightarrow$ make incompleteness seem **inevitable**

So this is what I've been working on. Now what is the reaction of the world to this work?! Well, I think it's fair to say that the only people who like what I'm doing are physicists! This is not surprising, because the idea came in a way from physics. I have a foreign idea called randomness that I'm bringing into logic, and logicians feel very uncomfortable with it. You know, the notion of program size, program-size complexity is like the idea of entropy in thermodynamics. So it turns out that physicists find this nice because they view it as ideas from their field invading logic. But logicians don't like this very much.

I think there may be political reasons, but I think there are also legitimate conceptual reasons, because these are ideas that are so foreign, the idea of randomness or of things that are true by accident is so foreign to a mathematician or a logician, that it's a nightmare! This is their worst nightmare come true! I think they would prefer not to think about it.

On the other hand, physicists think this is delightful! Because they remember well the crisis that they went through in the 1920's about randomness at the foundations of physics, and they say, it's not just us, we're not the only people who have randomness, pure math has it too, they're not any better than we are!

I'll give an example of the attitude of physicists to my theory. It just so happens that this week I found it by chance. There's an English magazine *New Scientist* that comes out every week; it's like an English version of *Scientific American*, except that it's a little livelier, it's a little more fun, and it comes out every week. And the current issue — the one that appeared February 26th, the next issue hasn't come out yet — of *New Scientist* has on its cover an article called "Random Reality". And if you open the issue and look at this article, it turns out to be an article about the work of two physicists, very speculative work. They're trying to get space and time, three or four dimensional spacetime, our world, to emerge from a random substratum underneath.

The reason that I mention this article is that these physicists say that their work was inspired by Gödel's and my work on the limits of logic; they're trying to absorb this stuff. They say that physicists were interested in Gödel's result, but they couldn't relate to it, it's not in terms that make sense to a physicist. But my work, they say, that makes sense to a physicist! It's not surprising: I got the idea by reading physics. So it makes sense to them because it's an idea that came from their field and is coming back to their field.

Actually, they don't use my definitions or my theorems at all, because I was asked to referee their paper, and I had to say that it really has nothing to do with me. My stuff is mentioned in the introduction because it helped to stimulate their work, but actually their work is in physics and has nothing to do with my area, which is algorithmic information theory.

But I think this is an interesting example of the fact that crazy ideas sometimes have unexpected consequences! As I said, formal systems did not succeed for reasoning, but they succeeded wonderfully for computation. So Hilbert is the most incredible success in the world, but as technology, not as epistemology.

And unexpectedly there are physicists who are interested in my notion of program-size complexity; they view it as another take on thermodynamical entropy. There's some work by real physicists on Maxwell's demon using my ideas; I mention this for those of you who have some physics background.

But I must say that philosophers have not picked up the ball. I think logicians hate my work, they detest it! And I'm like pornography, I'm sort of an unmentionable subject in the world of logic, because my results are so disgusting!

So this is my story! To end, let me quote from a posthumous collection of essays by Isaiah Berlin, *The Power of Ideas*, that was just published: "Over a hundred years ago, the German poet Heine warned the French not to underestimate the power of ideas: philosophical concepts nurtured in the stillness of a professor's study could destroy a civilization." So **beware of ideas**, I think it's really true.

Hilbert's idea of going to the limit, of complete formalization, which was for epistemological reasons, this was a philosophical controversy about the foundations of mathematics — are there foundations? And in a way this project failed, as I've explained, because of the work of Gödel and Turing. But here we are with these complete formalizations which are computer programming languages, they're everywhcrc! They pay my salary, they probably pay your salary... well, this is the School of Computer Science, it pays for all of this, right? Here we are!

So it worked! In another sense, it worked tremendously.

So I like to apologize in an aggressive way about my field. I like to say that my field has no applications, that the most interesting thing about the field of program-size complexity is that it has no applications, that it proves that it cannot be applied! Because you can't calculate the size of the smallest program. But that's what's fascinating about it, because it reveals limits to what we can know. That's why program-size complexity has epistemological significance.

More seriously, I think the moral of the story is that deep ideas don't have a spin-off in dollars right away, but sometimes they have vastly unexpected consequences. I never expected to see two physicists refer to my stuff the way they did in "Random Reality". So who knows!

It's true that the computer pays for our salaries but I think it's also true that there are a lot of fascinating impractical ideas out there. Someone told me at lunch today that an idea is so beautiful, it's got to be right. Those are the ideas to watch out for! Those are the dangerous ones, the ones that can transform our society. This little idea of a web, for example, of linking stuff into a web! Or the idea of having completely artificial languages, because then it becomes mechanical to see what they mean... Very dangerous ideas! Thanks very much!

Bibliography

1. G.J. Chaitin, *Algorithmic Information Theory,* Cambridge University Press, 1987.

2. G.J. Chaitin, *Information, Randomness & Incompleteness,* World Scientific, 1987.

3. G.J. Chaitin, *Information, Randomness & Incompleteness,* 2nd Ed., World Scientific, 1990.

4. G.J. Chaitin, *Information-Theoretic Incompleteness,* World Scientific, 1992.

5. G.J. Chaitin, *The Limits of Mathematics,* Springer-Verlag, 1998.

6. G.J. Chaitin, *The Unknowable,* Springer-Verlag, 1999.

7. I've recently finished programming my entire theory in LISP. This will eventually be my book *Exploring Randomness,* in preparation.

Metamathematics and the foundations of mathematics

This article discusses what can be proved about the foundations of mathematics using the notions of algorithm and information. The first part is retrospective, and presents a beautiful antique, Gödel's proof; the first modern incompleteness theorem, Turing's halting problem; and a piece of postmodern metamathematics, the halting probability Ω. The second part looks forward to the new century and discusses the convergence of theoretical physics and theoretical computer science and hopes for a theoretical biology, in which the notions of algorithm and information are again crucial.

PART I. THREE INCOMPLETENESS THEOREMS

In this article I'm going to concentrate on what we can **prove** about the foundations of mathematics using mathematical methods, in other words, on metamathematics. The current point of departure for metamathematics is that you're doing mathematics using an artificial language and you pick a **fixed** set of axioms and rules of inference (deduction rules), and everything is done so precisely that there is a proof-checking algorithm. I'll call such a formal system a formal axiomatic theory.

Then, as is pointed out in Turing's original paper (1936), and as was emphasized by Post in his *American Mathematical Society Bulletin* paper (1944), the set X of all theorems, consequences of the axioms, can be systematically generated by running through all possible proofs in size order and mechanically checking which ones are valid. This unending computation, which would be monumentally time-consuming, is sometimes jocularly referred to as the British Museum algorithm.

The size in bits $H(X)$ of the program that generates the set X of theorems—that's the program-size complexity of X—will play a crucial role below. Roughly speaking, it's the number of bits of axioms in the formal theory that we are considering. $H(X)$ will give us a way to measure the algorithmic complexity or the algorithmic information content of a formal axiomatic theory.

But first let's retrace history, starting with a beautiful antique, Gödel's incompleteness theorem, the very first incompleteness theorem.

- Alan Turing (1936), "On computable numbers, with an application to the entscheidungsproblem," *Proceedings of the London Mathematical Society*, ser. 2, vol. 42, pp. 230–265. Reprinted in Davis (1965), pp. 115–154.

- Emil Post (1944), "Recursively enumerable sets of positive integers and their decision problems," *American Mathematical Society Bulletin*, vol. 50, pp. 284–316. Reprinted in Davis (1965), pp. 304–337.

- Martin Davis (1965), *The Undecidable*, Raven Press.

A Beautiful Antique: Gödel's Proof (1931)

Let's fix our formal axiomatic theory as above and ask if "This statement is unprovable!" can be proven within our theory. "This statement is unprovable!" is provable if and only if it's false! "This statement is unprovable!" doesn't sound at all like a mathematical statement, but Gödel shows that it actually is.

Therefore formal axiomatic theories, if they only prove true theorems, are incomplete, because they do not prove all true statements. And so true and provable turn out to be rather different!

How does Gödel's proof work? Gödel cleverly constructs an arithmetical or number-theoretic assertion that refers to itself and its unprovability indirectly, via the (Gödel) numbers of the statements and proofs within the formal theory. In other words, he numbers all the statements and proofs within the formal axiomatic theory, and he can then construct a (very complicated) bona fide mathematical assertion that states that it itself is unprovable. The self-reference is indirect, since Gödel's self-referential statement cannot contain its own Gödel number.

Wonderful as it is, Gödel's proof does not involve three of the big ideas of the 20th century, **algorithm, information** and **randomness**. The first step in that direction was taken by Turing only five years later.

But before discussing Turing's work, what is an algorithm?

- Kurt Gödel (1931), "Über formal unentscheidbare Sätze der Principia Mathematica und verwandter Systeme I" ["On formally undecidable propositions of *Principia Mathematica* and related systems I"], *Monatshefte für Mathematik und Physik*, vol. 38, pp. 173–198. English translation in Davis (1965), pp. 4–38. [Very difficult to understand.]

- Ernest Nagel, James R. Newman (1958), *Gödel's Proof*, New York University Press. [A beautifully clear explanation.]

What is an Algorithm? [McCarthy (1962), Chaitin (1998, 1999, 2001)]

An algorithm is a mechanical procedure for calculating something, usually formulated in a programming language, for example LISP, which is a computable version of set theory!

In LISP, which is my favorite programming language, applying the function f to the operands x and y, $f(x, y)$, is written (f x y).

Programs and data in LISP are all S-expressions, which are lists with sublists, enclosed in parentheses and with successive elements separated by blanks. For example, (A BC (123 DD)) is an S-expression. The S in S-expression stands for "symbolic."

For example, let's define set membership in LISP.

```
(define (in-set? member set)
   (if (= () set) false
   (if (= member (head set)) true
       (in-set? member (tail set))
   ))
)
```

This defines (in-set? member set) to be false if the set is empty, true if the member is the first element in the set, and to recursively be (in-set? member [the rest of the set]) otherwise.

Let me explain some of this: (if x y z) yields/picks y or z depending on whether or not x is true. And (= x y) checks if $x = y$, yielding true or false.

Unfortunately, for historical reasons, head is actually written car and tail is actually written cdr.

Then

```
(in-set? (' y) (' (x y z)))
```

yields **true**, and

```
(in-set? (' q) (' (x y z)))
```

yields **false**.

Here ' or *quote* stops evaluation and contains unevaluated data.

In summary, a LISP program isn't a list of statements that you execute or run, it's an expression to be evaluated, and what it does, is it yields a value. In other words, LISP is a functional programming language, not an imperative programming language.

- John McCarthy et al. (1962), *LISP 1.5 Programmer's Manual,* Massachusetts Institute of Technology Press.

The First Modern Incompleteness Theorem: Turing's Halting Problem (1936)

At the beginning of his 1936 paper, Turing provides a mathematical definition of the notion of algorithm. He does this using an extremely primitive programming language (now called a Turing machine), not LISP as above, but it is nevertheless a decisive intellectual step forward into the computer age.[1] He then proves that there are things which cannot be computed. Turing does this by making brilliant use of Cantor's diagonal argument from set theory applied to the list of all computable real numbers. This gives Turing an uncomputable real number R^*, as we explain below, and a completely different source of incompleteness than the one discovered by Gödel in 1931. Real numbers are numbers like 3.1415926...

Imagine a numbered list of all possible computer programs for computing real numbers. That is, a list of all possible computer programs in some fixed language, ordered by size, and within programs of the same size, in some arbitrary alphabetical order. So $R(N)$ is the real number (if any) that is computed by the Nth program in the list ($N = 1, 2, 3, \ldots$).

Let $R(N, M)$ be the Mth digit after the decimal point of the Nth computable real, that is, the real $R(N)$ calculated by the Nth program. Define a

---

[1] In Chaitin (1999) I discuss the halting problem and Gödel's proof using LISP.

new real R^* whose Nth digit after the decimal point, $R^*(N)$, is 3 if $R(N, N)$ is not equal to 3, and otherwise is 2 (including the case that the Nth computer program never outputs an Nth digit). Then R^* is an uncomputable real, because it differs from the Nth computable real in the Nth digit. Therefore there cannot be any way to decide if the Nth computer program ever outputs an Nth digit, or we could actually compute R^*, which is impossible.

Corollary: No formal axiomatic theory can always enable you to prove whether or not the Nth computer program ever outputs an Nth digit, because otherwise you could run through all possible proofs in size order and compute R^*, which is impossible.

Note: Whether the Nth computer program ever outputs an Nth digit is a special case of the halting problem, which is the problem of determining whether or not a computer program ever halts (with no time limit). If a program does halt, you can eventually determine that, merely by running it. The real problem is to decide that a program will never halt, no matter how much time you give it.

Postmodern Metamathematics: The Halting Probability Ω [Chaitin (1975, 1987, 1998), Delahaye (2002)]

So that's how Turing brought the notion of **algorithm** into metamathematics, into the discussion about the foundations of mathematics. Now let's find yet another source of incompleteness, and let's bring into the discussion the notions of **information, randomness, complexity** and **irreducibility.** First we need to define a certain kind of computer, or, equivalently, to specify its binary machine language.

What is the "self-delimiting binary universal computer" U that we use below?

U's program is a finite bit string, and starts off with a prefix that is a LISP expression. The LISP expression prefix is converted into binary, yielding 8 bits per character, and it's followed by a special punctuation character, 8 more bits, and that is followed by raw binary data. The LISP prefix is evaluated or run and it will read the raw binary data in one bit at a time without ever being allowed to run off the end of the program.

In other words, the LISP prefix must ask for exactly the correct number of bits of raw binary data. If it requests another bit after reading the last one, this does not return a graceful "end of file" condition, it aborts the computation.

The fact that there is no punctuation marking the end of the raw binary data, which is also the end of the entire program, is what forces these machine-language programs to be self-delimiting. In other words, the end of the binary program is like a cliff, and the computer U must not fall off!

What is the halting probability Ω for U?

$$\Omega = \sum_{p \text{ halts when run on } U} 2^{-(\text{the number of bits in } p)}.$$

So if the program p halts and is K bits long, that contributes $1/2^K$ to the halting probability Ω.

This halting probability can be defined in such a manner that it includes binary programs p of every possible size precisely because these p must be self-delimiting. That is to say, precisely because U must decide **by itself** where to stop reading the program p. U must not overshoot, it cannot fall off the cliff, it must read precisely up to the last bit of p, but not beyond it.

Knowing the first N bits of the base-two representation of the real number Ω, which is a probability and therefore between zero and one, answers the halting problem for all programs up to N bits in size. So if you knew the first N bits of Ω after the decimal or the binary point, that would theoretically enable you to decide whether or not each binary computer program p up to N bits in size halts when run on U.

Would this be useful? Yes, indeed! Let me give an example showing just how very useful it would be.

Let's consider the Riemann hypothesis, a famous mathematical conjecture that is still open, still unresolved. There is a Riemann-hypothesis testing program that systematically searches for counter-examples and that halts if and only if it finds one and the Riemann hypothesis is false. The size in bits of this program is the program-size complexity of testing the Riemann hypothesis this way. And if this program is $H(\text{Riemann})$ bits long, knowing that many initial bits of Ω would settle the Riemann hypothesis! It would enable you to tell whether or not the Riemann hypothesis is true.

Unfortunately the method required to do this, while theoretically sound, is totally impractical. The computation required, while finite, is much, much too long to actually carry out. The time needed grows much, much faster than exponentially in $H(\text{Riemann})$, the number of bits of Ω that we are given.

In fact, it grows as the time required to simultaneously run all programs on U up to H(Riemann) bits in size until all the programs that will ever halt have done so. More precisely, you have to run enough programs for enough time to get the first H(Riemann) bits of Ω right, which because of carries from bits that are further out may actually involve programs that are more than H(Riemann) bits long.

And precisely because the bits of Ω are so useful, it turns out that they are irreducible mathematical information, that they cannot be derived or deduced from any simpler principles. More precisely, we have the following incompleteness result: You need an N-bit formal axiomatic theory (that is, one that has an N-bit algorithm to generate all the theorems) in order to be able to determine the first N bits of Ω, or, indeed, the values and positions of any N bits of Ω.

Actually, what I show in Chaitin (1998) is that an N-bit theory can't determine more than $N + c$ bits of Ω, where the constant c is 15328.

Let's restate this. Consider a formal axiomatic theory with the set of theorems X and with algorithmic complexity or algorithmic information content $H(X)$. Then if a statement such as

"The 99th bit of Ω is 0."
"The 37th bit of Ω is 1."

determining the value of a particular bit of Ω in a particular place, is in X only if it's true, then there are at most $H(X) + c$ such theorems in X. In other words, X enables us to determine at most $H(X) + c$ bits of Ω.

We can also describe this irreducibility non-technically, but very forcefully, as follows: Whether each bit of Ω is a 0 or a 1 is a mathematical fact that is true for no reason, it's true by accident!

What is Algorithmic Information? [Chaitin (1975, 1987, 2001), Calude (2002)]

Ω is actually just one piece of my algorithmic information theory (AIT), it's the jewel that I discovered while I was developing AIT. Let me now give some highlights of AIT.

What else can we do using the computer U that we used to define Ω? Well, you should look at the size of programs for U. U is the yardstick you use to measure algorithmic information. And the unit of algorithmic information is the 0/1 bit.

You define the absolute algorithmic *information content* $H(X)$ of an object (actually, of a LISP S-expression) X to be the size in bits of the smallest program for U to compute X. The *joint* information content $H(X,Y)$ is defined to be the size in bits of the smallest program for U to compute the pair X, Y. (Note that the pair X, Y is actually $(X\ Y)$ in LISP.) The *relative* information content $H(X|Y)$ is defined to be the size in bits of the smallest program for U that computes X from a minimum-size program for Y, not from Y directly.

And the complexity $H(X)$ of a formal axiomatic theory with theorem set X is also defined using the computer U. (I glossed over this point before.) $H(X)$ is defined to be the size in bits of the smallest program that makes U generate the set of theorems X. Note that this is an endless computation. You may think of $H(X)$ as the number of bits of information in the most concise or the most elegant set of axioms that yields the set of theorems X.

Now here are some of the theorems that you can prove about these concepts.

First of all, let's consider an N-bit string X. $H(X)$ is usually close to $N + H(N)$, which is approximately $N + \log_2 N$. Bit strings X for which this is the case are said to be **algorithmically random,** they have the highest possible information content.

On the other hand, an infinite sequence of bits X is defined to be algorithmically random if and only if there is a constant c such that H(the first N bits of X) is greater than $N - c$ for all N. And, crucial point, the base-two representation for Ω satisfies this definition of algorithmic randomness, which is one of the reasons that Ω is so interesting.

Algorithmic information is (sub)additive:

$$H(X,Y) \leq H(X) + H(Y) + c.$$

And the *mutual* information content $H(X : Y)$ is defined to be the extent to which computing two objects together is better than computing them separately:

$$H(X : Y) = H(X) + H(Y) - H(X,Y).$$

X and Y are said to be *algorithmically independent* if their mutual information is small compared with their individual information contents, so that

$$H(X,Y) \approx H(X) + H(Y).$$

Finally, here are some subtle results that relate mutual and relative information:

$$H(X,Y) = H(X) + H(Y|X) + O(1),$$

$$H(X : Y) = H(X) - H(X|Y) + O(1),$$

$$H(X : Y) = H(Y) - H(Y|X) + O(1).$$

Here l.h.s. = r.h.s. + $O(1)$ means that the difference between the left-hand side and the right-hand side of the equation is bounded, it's at most a fixed number of bits. Thus the mutual information is also the extent to which knowing one of a pair helps you to know the other.

These results are quoted here in order to show that Ω isn't isolated, it's part of an elegant theory of algorithmic information and randomness, a theory of the program-size complexity for U.

Now let me tell you what I think is the significance of these incompleteness results and of Ω.

Is Mathematics Quasi-Empirical?

That is, is mathematics more like physics than mathematicians would like to admit? I think so!

I think that incompleteness cannot be dismissed and that mathematicians should occasionally be willing to add new axioms that are justified by experience, experimentally, pragmatically, but are not at all self-evident.[2] Sometimes to prove more, you need to assume more, to add new axioms! That's what my information-theoretic approach to incompleteness suggests to me.

Of course, at this point, at the juncture of the 20th and the 21st centuries, this is highly controversial. It goes against the current paradigm of what mathematics is and how mathematics should be done, it goes against the current paradigm of the nature of the mathematical enterprise. But my hope is that the 21st century will eventually decide that adding new axioms is not at all controversial, that it's obviously the right thing to do! However this radical paradigm shift may take many years of discussion and thought to be accepted, if indeed this ever occurs.

For further discussion of this quasi-empirical, experimental mathematics viewpoint, see Chaitin (1998, 1999, 2002), Tymoczko (1998), Borwein (2002).

---

[2]In my opinion $\mathbf{P} \neq \mathbf{NP}$ is a good example of such a new axiom.

For superb histories of many aspects of 20th century thought regarding the foundations of mathematics that we have not touched upon here, see Grattan-Guinness (2000), Tasić (2001).

General Bibliography for Part I

- Jonathan Borwein, David Bailey (2002?), *Mathematics by Experiment*, A. K. Peters. [In preparation.]

- Cristian Calude (2002?), *Information and Randomness*, Springer-Verlag. [In preparation.]

- Gregory Chaitin (1975), "A theory of program size formally identical to information theory," *Association for Computing Machinery Journal*, vol. 22, pp. 329–340.

- Gregory Chaitin (1987), *Algorithmic Information Theory*, Cambridge University Press.

- Gregory Chaitin (1998, 1999, 2001, 2002), *The Limits of Mathematics, The Unknowable, Exploring Randomness, Conversations with a Mathematician*, Springer-Verlag.

- Jean-Paul Delahaye (2002), "Les nombres oméga," *Pour la Science*, mai 2002, pp. 98–103.

- Ivor Grattan-Guinness (2000), *The Search for Mathematical Roots, 1870-1940*, Princeton University Press.

- Vladimir Tasić (2001), *Mathematics and the Roots of Postmodern Thought*, Oxford University Press.

- Thomas Tymoczko (1998), *New Directions in the Philosophy of Mathematics*, Princeton University Press.

PART II. FUTURE PERSPECTIVES

Where is Metamathematics Going?

We need a dynamic, not a static metamathematics, one that deals with the evolution of new mathematical concepts and theories. Where do new mathematical ideas come from? Classical metamathematics with its incompleteness theorems deals with a static view of mathematics, it considers a **fixed** formal axiomatic system. But mathematics is constantly evolving and changing! Can we explain how this happens? What we really need now is a new, optimistic **dynamic** metamathematics, not the old, pessimistic **static** metamathematics.

In my opinion the following line of research is relevant and should not have been abandoned:

- Douglas Lenat (1984), "Automated theory formation in mathematics," in W. W. Bledsoe, D. W. Loveland, *Automated Theorem Proving: After 25 Years*, American Mathematical Society, pp. 287–314.

Where is Mathematics Going?

Will mathematics become more like biology, more complicated, less elegant, with more and more complicated theorems and longer and longer proofs?

Will mathematics become more like physics, more experimental, more quasi-empirical, with fewer proofs?

For a longer discussion of this, see the chapter on mathematics in the third millennium in Chaitin (1999).

What is a Universal Turing Machine (UTM)?

As physicists have become more and more interested in complex systems, the notion of algorithm has become increasing important, together with the idea that what physical systems actually do is computation. In other words, due to complex systems, physicists have begun to consider the notion of algorithm as physics. And the UTM now begins to emerge as a fundamental **physical** concept, not just a mathematical concept [Deutsch (1997), Wolfram (2002)].

To a mathematician a UTM is a formal, artificial, unambiguous language for formulating algorithms, a language that can be interpreted mechanically, and which enables you to specify any algorithm, all possible algorithms. That's why it's called "universal."

What is a UTM to a physicist? Well, it's a physical system whose repertoire of potential behavior is extremely rich, in fact maximally rich, universal, because it can carry out any computation and it can simulate the behavior of any other physical system.

These are two sides of a single coin.

And, in a sense, all of this was anticipated by Turing in 1936 when he used the word **machine.** Also the "halting problem" almost sounds like a problem in physics. It sounds very down-to-earth and concrete. It creates a mental image that is more physical than mathematical, it sounds like you are trying to stop a runaway locomotive!

After all, what you can compute depends on the laws of physics. A different universe might have different computers. So, in a way, Turing's 1936 paper was a physics paper!

- David Deutsch (1997), *The Fabric of Reality,* Penguin.

- Stephen Wolfram (2002), *A New Kind of Science,* Wolfram Media.

Convergence of Theoretical Physics and Theoretical Computer Science

The fact that "UTM" is now in the mental tool kit of physicists as well as mathematicians is just one symptom. What we are witnessing now is broader than that, it's actually the beginning of an amazing convergence of theoretical physics and theoretical computer science, which would have seemed inconceivable just a few years ago. There are many lines of research, many threads, that now go in that direction. Let me indicate some of these here.

It is sometimes useful to think of physical systems as performing algorithms, and of the entire universe as a single giant computer. Edward Fredkin was one of the earliest proponents of this view. See Wright (1988).

There is an increasing amount of work by physicists that suggests that it is fertile to view physical systems as information-processing systems, and that studies how physical systems process information. The extremely popular field of research of quantum computation and quantum information [Nielsen (2000)] certainly follows this paradigm.

And there are also suggestions from black hole thermodynamics and quantum mechanics that the physical universe may actually be discrete, not continuous, and that the maximum amount of information contained in a physical system is actually finite. A leading researcher in this area is Jacob Bekenstein, and for more on this topic, see the chapter on the holographic principle in Smolin (2001).

Wolfram (2002) is a treasure-trove of simple combinatorial (symbolic, discrete, non-numerical) algorithms with extremely rich behavior (in fact, universal behavior, equivalent to a UTM, a universal Turing machine, in other words, that can perform an arbitrary computation and simulate an arbitrary physical system). These are superb building blocks that God might well have used in building the universe! Here I refer to high-energy or particle physics. Time will tell—we will see! Wolfram also supports, and himself employs, an experimental, quasi-empirical approach to doing mathematics.

Let me also cite the physicist who might be considered the inventor of quantum computing, and also cite a journalist. See the chapter on "Universality and the limits of computation" in Deutsch (1998), and Siegfried (2000) on the physics of information and the seminal ideas of Rolf Landauer.

I should mention that my own work on AIT in a sense belongs to this school; it can be viewed as an application of the physical notion of entropy (or disorder) to metamathematics. In other words, my work on Ω in a sense treats formal axiomatic theories as if they were heat engines. That is how I show that Ω is irreducible, using general, "thermodynamical" arguments on the limitations of formal axiomatic theories.

Another example of ideas from physics that are invading computer science are the phase changes that mark a sharp transition from a regime in which an algorithm is fast to a regime in which the algorithm is extremely slow, for instance the situation described in Hayes (2002).

- Robert Wright (1988), *Three Scientists and Their Gods,* Times Books. [On Edward Fredkin.]

- Michael Nielsen, Isaac Chuang (2000), *Quantum Computation and Quantum Information,* Cambridge University Press.

- Tom Siegfried (2000), *The Bit and the Pendulum,* Wiley. [For the work of Rolf Landauer.]

- Lee Smolin (2001), *Three Roads to Quantum Gravity,* Basic Books.

- Brian Hayes (2002), "The easiest hard problem," *American Scientist,* vol. 90, pp. 113–117.

To a Theoretical Biology

What is life? Can there be a general, abstract mathematical theory of the origin and the evolution of the complexity of life?

Ergodic theory says that things get washed out and less interesting as time passes. The theory I want would show that interesting things (life! organisms! us!) emerge and evolve spontaneously, and that things get more and more interesting, not less and less interesting.

My old attempt at a theory [Chaitin (1970, 1979)] considered cellular automata and proposed using mutual algorithmic information to distinguish a living organism from its environment. My idea was that the parts of an organism have high mutual information.

Wolfram (2002) sustains the thesis that life is not unusual. He claims that there is no essential difference between us and any other universal Turing machine. Furthermore, according to Wolfram, most non-trivial physical and combinatorial systems are universal Turing machines, UTM's. (A UTM is a physical system whose behavior is as rich as possible because it is a general-purpose computer that can perform an arbitrary computation, in other words, that can simulate any algorithm and any other physical system.) Therefore, according to Wolfram, there is nothing to Darwin, nothing to understand. The evolution of life is a non-issue. According to Wolfram you get there, you get life, right away, all at once.

Wolfram's thesis, while interesting, is not, I believe, the entire story. Universal Turing machines, computation, may be ubiquitous in nature, but the amount of software a UTM has is not taken into account by Wolfram. And that, I believe, is what is actually evolving! After all, DNA is essentially digital software, and we have much more DNA than viruses and bacteria. Our program-size complexity is higher, H(human) is much greater than H(bacteria).

This suggests to me a new toy model of evolution very different from the cellular automata model that I originally considered. My new idea is to model life, an ecology, as a collection of UTM's, and to study how their software evolves in complexity. The problem is how to model the environment, more precisely, the interactions of organisms with each other and with their environment. Let me emphasize that in such a model the problem is not to distinguish an organism from its environment, which before I attempted to do using mutual information, it is to model interactions, so that the organisms are not like Leibniz's windowless monads! And then of course to prove that the software complexity will evolve. . .

For more on mutual information, see Chaitin (2001). For further discussion of my hopes for a theoretical biology, and for my thoughts on biology in general, see Chaitin (2002). For von Neumann's seminal work in this area, see von Neumann (1966). Two recent books on biology that particularly impressed me are Maynard Smith (1999), and Kay (2000).

- John von Neumann (1966), *Theory of Self-Reproducing Automata,* edited and completed by Arthur Burks, University of Illinois Press.

- Gregory Chaitin (1970, 1979), "To a mathematical definition of 'life'," *ACM SICACT News,* January 1970, pp. 12–18; "Toward a mathematical definition of 'life'," in R. D. Levine, M. Tribus, *The Maximum Entropy Formalism,* Massachusetts Institute of Technology Press, pp. 477–498.

- John Maynard Smith, Eörs Szathmáry (1999), *The Origins of Life,* Oxford University Press.

- Lily Kay (2000), *Who Wrote the Book of Life?,* Stanford University Press. [On information theory and molecular biology.]

To a Theoretical Psychology

What is psychological information? What is thinking? What is the soul? What is intelligence? Is it some kind of information-processing capability? How can we measure it? How can we simulate it? (That's called AI, artificial intelligence.) And where do new ideas come from? Ideas in general, not just mathematical ideas.

See Nørretranders (1998) for some interesting discussions of precisely these questions.

However, again, I don't just want an interesting discussion, I want a mathematical theory with beautiful theorems and proofs! Human intelligence may just be a very complicated piece of engineering, or there may be some profound, basic, as yet unknown concepts at play, and a fundamental theory about them. And these two possibilities are not mutually exclusive. Time will tell!

- Tor Nørretranders (1998), *The User Illusion,* Viking. [On information theory and psychology.]

To the Future!

I trust that a hundred years from now mathematicians will be able to look back on our current mathematics and metamathematics the way we regard the mathematics of 1900—with a justified feeling of superiority! I hope that they will wonder, how we could have been so blind, to miss the simple, wonderful, beautiful new theories that were just around the corner? I hope that they will ask themselves, how come we didn't see all those beautiful new ideas, when we were almost close enough to touch them and to taste them!

Paradoxes of randomness

*I'll discuss how Gödel's paradox "This statement is false/unprovable" yields his famous result on the limits of axiomatic reasoning. I'll contrast that with my work, which is based on the paradox of "The first uninteresting positive whole number," which is itself a rather interesting number, since it is precisely the first uninteresting number. This leads to my first result on the limits of axiomatic reasoning, namely that most numbers are uninteresting or random, but we can never be sure, we can never prove it, in individual cases. And these ideas culminate in my discovery that some mathematical facts are true for no reason, they are true by accident, or at random. In other words, God not only plays dice in physics, but even in pure mathematics, in logic, in the world of pure reason. Sometimes mathematical truth is completely random and has no structure or pattern that we will ever be able to understand. It is **not** the case that simple clear questions have simple clear answers, not even in the world of pure ideas, and much less so in the messy real world of everyday life. [This talk was given Monday 13 May 2002 at Monash University in Melbourne, Australia, and previously to summer visitors at the IBM Watson Research Center in 2001. There are no section titles; the displayed material is what I wrote on the whiteboard as I spoke.]*

When I was a small child I was fascinated by magic stories, because they postulate a hidden reality behind the world of everyday appearances. Later I switched to relativity, quantum mechanics, astronomy and cosmology, which also seemed quite magical and transcend everyday life. And I learned that physics says that the ultimate nature of reality is mathematical, that math is more real than the world of everyday appearances. But then I was surprised to learn of an amazing, mysterious piece of work by Kurt Gödel that pulled the rug out from under mathematical reality! How could this be?! How could Gödel show that math has limitations? How could Gödel use mathematical reasoning to show that mathematical reasoning is in trouble?!

Applying mathematical methods to study the power of mathematics is called meta-mathematics, and this field was created by David Hilbert about a century ago. He did this by proposing that math could be done using a completely artificial formal language in which you specify the rules of the game so precisely that there is a mechanical procedure to decide if a proof is correct or not. A formal axiomatic theory of the kind that Hilbert proposed would consist of axioms and rules of inference with an artificial grammar and would use symbolic logic to fill in all the steps, so that it becomes completely mechanical to apply the rules of inference to the axioms in every possible way and systematically deduce all the logical consequences. These are called the theorems of the formal theory.

You see, once you do this, you can forget that your formal theory has any meaning and study it from the outside as if it were a meaningless game for generating strings of symbols, the theorems. So that's how you can use mathematical methods to study the power of mathematics, if you can formulate mathematics as a formal axiomatic theory in Hilbert's sense. And Hilbert in fact thought that all of mathematics could be put into one of his formal axiomatic theories, by making explicit all the axioms or self-evident truths and all the methods of reasoning that are employed in mathematics.

In fact, Zermelo-Fraenkel set theory with the axiom of choice, ZFC, uses first-order logic and does this pretty well. And you can see some interesting work on this by Jacob T. Schwartz at his website at `http://www.settheory.com`.

But then in 1931 Kurt Gödel showed that it couldn't be done, that no formal axiomatic theory could contain all of mathematical truth, that they were all incomplete. And this exploded the normal Platonic view of what math is all about.

How did Gödel do this? How can mathematics prove that mathematics has limitations? How can you use reasoning to show that reasoning has limitations?

How does Gödel show that reasoning has limits? The way he does it is he uses this paradox:

"This statement is false!"

You have a statement which says of itself that it's false. Or it says

"I'm lying!"

"I'm lying" doesn't sound too bad! But "the statement I'm making now is a lie, what I'm saying right now, this very statement, is a lie," that sounds worse, doesn't it? This is an old paradox that actually goes back to the ancient Greeks, it's the paradox of the liar, and it's also called the Epimenides paradox, that's what you call it if you're a student of ancient Greece.

And looking at it like this, it doesn't seem something serious. I didn't take this seriously. You know, so what! Why should anybody pay any attention to this? Well, Gödel was smart, Gödel showed why this was important. And Gödel changed the paradox, and got a theorem instead of a paradox. So how did he do it? Well, what he did is he made a statement that says of itself,

"This statement is unprovable!"

Now that's a big, big difference, and it totally transforms a game with words, a situation where it's very hard to analyze what's going on. Consider

"This statement is false!"

Is it true, is it false? In either case, whatever you assume, you get into trouble, the opposite has got to be the case. Why? Because if it's true that the statement is false, then it's false. And if it's false that the statement is false, then it's true.

But with

"This statement is unprovable!"

you get a theorem out, you don't get a paradox, you don't get a contradiction. Why? Well, there are two possibilities. With

"This statement is false!"

you can assume it's true, or you can assume it's false. And in each case, it turns out that the opposite is then the case. But with

"This statement is unprovable!"

the two possibilities that you have to consider are different. The two cases are: it's provable, it's unprovable.

So if it's provable, and the statement says it's unprovable, you've got a problem, you're proving something that's false, right? So that would be very embarrassing, and you generally assume by hypothesis that this cannot be the case, because it would really be too awful if mathematics were like that. If mathematics can prove things that are false, then mathematics is in trouble, it's a game that doesn't work, it's totally useless.

So let's assume that mathematics does work. So the other possibility is that this statement

"This statement is unprovable!"

is unprovable, that's the other alternative. Now the statement is unprovable, and the statement says of itself that it's unprovable. Well then it's true, because what it says corresponds to reality. And then there's a **hole** in mathematics, mathematics is "incomplete," because you've got a true statement that you can't prove. The reason that you have this hole is because the alternative is even worse, the alternative is that you're proving something that's false.

The argument that I've just sketched is not a mathematical proof, let me hasten to say that for those of you who are mathematicians and are beginning to feel horrified that I'm doing everything so loosely. This is just the basic idea. And as you can imagine, it takes some cleverness to make a statement in mathematics that says of itself that it's unprovable. You know, you don't normally have pronouns in mathematics, you have to have an indirect way to make a statement refer to itself. It was a very, very clever piece of work, and this was done by Gödel in 1931.

1931

The only problem with Gödel's proof is that I didn't like it, it seemed strange to me, it seemed beside the point, I thought there had to be a better, deeper reason for incompleteness. So I came up with a different approach, another way of doing things. I found a different source for incompleteness.

Now let me tell you my approach. My approach starts off like this... I'll give you two versions, a simplified version, and a slightly less-of-a-lie version.

The simplified version is, you divide all numbers into two classes, you think of whether numbers are interesting or uninteresting, and I'm talking about whole numbers, positive integers,

$$1, 2, 3, 4, 5, \ldots$$

That's the world I'm in, and you talk about whether they're interesting or uninteresting.

Un/Interesting

Somehow you separate them into those that are interesting, and those that are uninteresting, okay? I won't tell you how. Later I'll give you more of a clue, but for now let's just keep it like that.

So, the idea is, then, if somehow you can separate all of the positive integers, the whole numbers, 1, 2, 3, 4, 5, into ones that are interesting and ones that are uninteresting, you know, each number is either interesting or uninteresting, then think about the following whole number, the following positive integer:

"The first uninteresting positive integer"

Now if you think about this number for a while, it's precisely what? You start off with 1, you ask is it interesting or not. If it's interesting, you keep going. Then you look and see if 2 is interesting or not, and precisely when you get to the first uninteresting positive integer, you stop.

But wait a second, isn't that sort of an interesting fact about this positive integer, that it's **precisely** the first uninteresting positive integer?! I mean, it stands out that way, doesn't it? It's sort of an interesting thing about it, the fact that it happens to be precisely the smallest positive integer that's uninteresting! So that begins to give you an idea that there's a problem, that there's a serious problem with this notion of interesting versus uninteresting.

Interestingly enough, last week I gave this talk at the University of Auckland in New Zealand, and Prof. Garry Tee showed me the *Penguin Dictionary of Curious and Interesting Numbers* by David Wells that was published in Great Britain in 1986. And I'll read what it says on page 120: "**39**—This appears to be the first uninteresting number, which of course makes it an especially interesting number, because it is the smallest number to have the property of being uninteresting." So I guess if you read his dictionary you will find that the entries for the positive integers 1 through 38 indicate that each of them is interesting for some reason!

And now you get into a problem with mathematical proof. Because let's assume that somehow you can use mathematics to **prove** whether a number is interesting or uninteresting. First you've got to give a rigorous definition of this concept, and later I'll explain how that goes. If you can do that, and if you can also prove whether particular positive integers are interesting or uninteresting, you get into trouble. Why? Well, just think about the first positive integer that you can **prove** is uninteresting.

"The first **provably** uninteresting positive integer"

We're in trouble, because the fact that it's precisely the first positive integer that you can **prove** is uninteresting, is a very interesting thing about it! So if there cannot be a first positive integer that you can prove is uninteresting, the conclusion is that you can **never** prove that particular positive integers are uninteresting. Because if you could do that, the first one would *ipso facto* be interesting!

But I should explain that when I talk about the first provably uninteresting positive integer I **don't** mean the smallest one, I mean the first one that you find when you systematically run through all possible proofs and generate all the theorems of your formal axiomatic theory. I should also add that when you carefully work out all the details, it turns out that you might be able to prove that a number is uninteresting, but not if its base-two representation is substantially larger than the number of bits in the program for systematically generating all the theorems of your formal axiomatic theory. So you can only prove that a finite number of positive integers are uninteresting.

So that's the general idea. But this paradox of whether you can classify whole numbers into uninteresting or interesting ones, that's just a simplified version. Hopefully it's more understandable than what I actually worked with, which is something called the Berry paradox. And what's the Berry paradox?

Berry Paradox

I showed you the paradox of the liar, "This statement is false, I'm lying, what I'm saying right now is a lie, it's false." The Berry paradox talks about

"The first positive integer that can't be named
in less than a billion words"

Or you can make it bytes, characters, whatever, you know, some unit of measure of the size of a piece of text:

Berry Paradox
"The first positive integer that can't be named
in less than a billion words/bytes/characters"

So you use texts in English to name a positive integer. And if you use texts up to a billion words in length, there are only a finite number of them, since there are only a finite number of words in English. Actually we're simplifying, English is constantly changing. But let's assume English is fixed and you don't add words and a dictionary has a finite size. So there are only a finite number of words in English, and therefore if you consider all possible texts with up to a billion words, there are a lot of them, but it's only a finite number, as mathematicians say jokingly in their in-house jargon.

And most texts in English don't name positive integers, you know, they're novels, or they're actually nonsense, gibberish. But if you go through all possible texts of up to a billion words, and there's only a finite list of them, every possible way of using an English text that size to name a number will be there somewhere. And there are only a finite number of numbers that you can name with this finite number of texts, because to name a number means to pick out one specific number, to refer to precisely one of them. But there are an infinite number of positive integers. So most positive integers, almost all of them, require more than a billion words, or any fixed number of words. So just take the first one. Since almost all of them need more than a billion words to be named, just pick the first one.

So this number is there. The only problem is, I just named it in much less than a billion words, even with all the explanation! [Laughter] Thanks for smiling and laughing! If nobody smiles or laughs, it means that I didn't explain it well! On a good day everyone laughs!

So there's a problem with this notion of naming, and this is called the Berry paradox. And if you think that the paradox of the liar, "this statement is false," or "what I'm saying now is a lie," is something that you shouldn't take too seriously, well, the Berry paradox was taken even less seriously. I took it seriously though, because the idea I extracted from it is the idea of looking at the size of computer programs, which I call program-size complexity.

Program-Size Complexity

For me the central idea of this paradox is how big a text does it take to name something. And the paradox originally talks about English, but that's much too vague! So to make this into mathematics instead of just being a joke, you have to give a rigorous definition of what language you're using and how something can name something else. So what I do is I pick a computer-programming language instead of using English or any real language, any natural language, I pick a computer-programming language instead. And then what does it mean, how do you name an integer? Well, you name an integer by giving a way to calculate it. A program names an integer if its output is that integer, you know, it outputs that integer, just one, and then it stops. So that's how you name an integer using a program.

And then what about looking at the size of a text measured in billions of words? Well, you don't want to talk about words, that's not a convenient measure of software size. People in fact in practice use megabytes of code, but since I'm a theoretician I use bits. You know, it's just a multiplicative constant conversion factor! In biology the unit is kilobases, right? So every field has its way of measuring information.

Okay, so what does it mean then for a number to be interesting or uninteresting, now that I'm giving you a better idea of what I'm talking about. Well, interesting means it stands out some way from the herd, and uninteresting means it can't be distinguished really, it's sort of an average, typical number, one that isn't worth a second glance. So how do you define that mathematically using this notion of the size of computer programs? Well, it's very simple: a number is uninteresting or algorithmically random or irreducible or incompressible if there's no way to name it that's more concise than just writing out the number directly. That's the idea.

In other words, if the most concise computer program for calculating a number just says to print 123796402, in that case, if that's the best you can do, then that number is uninteresting. And that's typically what happens. On the other hand, if there is a small, concise computer program that calculates the number, that's atypical, that means that it has some quality or characteristic that enables you to pick it out and to compress it into a smaller algorithmic description. So that's unusual, that's an interesting number.

Once you set up this theory properly, it turns out that most numbers, the great majority of positive integers, are uninteresting. You can prove that as a theorem. It's not a hard theorem, it's a counting argument. There can't be a lot of interesting numbers, because there aren't enough concise programs. You know, there are a lot of positive integers, and if you look at programs with the same size in bits, there are only about as many programs of the same size as there are integers, and if the programs have to be smaller, then there just aren't enough of them to name all of those different positive integers.

So it's very easy to show that **the vast majority** of positive integers cannot be named substantially more concisely than by just exhibiting them directly. Then my key result becomes, that in fact you can never prove it, not in **individual** cases! Even though most positive integers are uninteresting in this precise mathematical sense, you can never be sure, you can never prove it—although there may be a finite number of exceptions. But you can only prove it in a small number of cases. So most positive integers are uninteresting or algorithmically incompressible, but you can almost never be sure in individual cases, even though it's overwhelmingly likely.

That's the kind of "incompleteness result" I get. (That's what you call a result stating that you can't prove something that's true.) And my incompleteness result has a very different flavor than Gödel's incompleteness result, and it leads in a totally different direction. Fortunately for me, everyone liked the liar paradox, but nobody took the Berry paradox really seriously!

Let me give you another version of this result. Let's pick a computer programming language, and I'll say that a computer program is **elegant** if no program that is smaller than it is produces the same output that it does. Then you can't prove that a program is elegant if it's substantially larger than the algorithm for generating all the theorems of the formal axiomatic theory that you are using, if that's written in that same computer programming language. Why?

Well, start generating all the theorems until you find the first one that proves that a particular computer program that is larger than that is elegant. That is, find the first provably elegant program that's larger than the program in the same language for generating all the theorems. Then run it, and its output will be your output.

I've just described a program that produces the same output as a provably elegant program, but that's smaller than it is, which is impossible! This contradiction shows that you can only prove that a finite number of programs are elegant, if you are using a fixed formal axiomatic theory.

By the way, this implies that you can't always prove whether or not a program halts, because if you could do that then it would be easy to determine whether or not a program is elegant. So I'm really giving you an information-theoretic perspective on what's called Turing's halting problem, I'm connecting that with the idea of algorithmic information and with program-size complexity.

I published an article about all of this in *Scientific American* in 1975, it was called "Randomness and mathematical proof," and just before that I called Gödel on the phone to tell him about it, that was in 1974.

I was working for IBM in Buenos Aires at the time, and I was visiting the

IBM Watson Research Center in New York—that was before I joined IBM Research permanently. And just before I had to go back to Buenos Aires I called Gödel on the phone at the Princeton Institute for Advanced Study and I said, "I'm fascinated by your work on incompleteness, and I have a different approach, using the Berry paradox instead of the paradox of the liar, and I'd really like to meet you and tell you about it and get your reaction." And he said, "It doesn't make any difference which paradox you use!" (And his 1931 paper said that too.) I answered, "Yes, but this suggests to me a new information-theoretic view of incompleteness that I'd very much like to tell you about." He said, "Well, send me a paper on this subject and call me back, and I'll see if I give you an appointment."

I had one of my first papers then, actually it was the proofs of one of my first papers on the subject. It was my 1974 *IEEE Information Theory Transactions* paper; it's reprinted in Tymoczko, *New Directions in the Philosophy of Mathematics*. And I mailed it to Gödel. And I called back. And incredibly enough, he made a small technical remark, and he gave me an appointment. I was delighted, you can imagine, my hero, Kurt Gödel! And the great day arrives, and I'm in my office in the Watson Research Center at Yorktown Heights, NY, and it was April 1974, spring. In fact, it was the week before Easter. And I didn't have a car. I was coming from Buenos Aires, I was staying at the YMCA in White Plains, but I figured out how to get to Princeton, New Jersey by train. You know, I'd take the train into New York City and then out to Princeton. It would only take me three hours, probably, to do it!

So I'm in my office, ready to go, almost, and the phone rings. And I forgot to tell you, even though it was the week before Easter, it had snowed. It wasn't a whole lot of snow; you know, nothing would stop me from visiting my hero Gödel at Princeton. So anyway, the phone rings, and it's Gödel's secretary, and she says, "Prof. Gödel is extremely careful about his health, and because it's snowed, he's not going to be coming in to the Institute today, so your appointment is canceled!"

And as it happened, that was just two days before I had to take a plane back to Buenos Aires from New York. So I didn't get to meet Gödel! This is one of the stories that I put in my book *Conversations with a Mathematician*.

So all it takes is a new idea! And the new idea was waiting there for anybody to grab it. The other thing you have to do when you have a new idea is, don't give up too soon. As George Polya put it in his book *How to Solve It*, theorems are like mushrooms, usually where there's one, others will pop up! In other words, another way to put it, is that usually the difference between a professional, expert mathematician with lots of experience and a young, neophyte mathematician is not that the older mathematician has more ideas. In fact, the opposite is usually the case. It's usually the kids that have all the fresh ideas! It's that the professional knows how to take more advantage of the ideas he has. And one of the things you do, is you don't give up on an idea until you get all the milk, all the juice out of it!

So what I'm trying to lead up to is that even though I had an article in *Scientific American* in 1975 about the result I just told you, that most numbers are random, algorithmically random, but you can never prove it, I didn't give up, I kept thinking about it. And sure enough, it turned out that there was another major result there, that I described in my article in *Scientific American* in 1988. Let me try to give you the general idea.

The conclusion is that

Some mathematical facts are true for no reason, they're true by accident!

Let me just explain what this means, and then I'll try to give an idea of how I arrived at this surprising conclusion. The normal idea of mathematics is that if something is true it's true for a reason, right? The reason something is true is called a proof. And a simple version of what mathematicians do for a living is they find proofs, they find the reason that something is true.

Okay, what I was able to find, or construct, is a funny area of pure mathematics where things are true for no reason, they're true by accident. And that's why you can never find out what's going on, you can never prove what's going on. More precisely, what I found in pure mathematics is a way to model or imitate, independent tosses of a fair coin. It's a place where God plays dice with mathematical truth. It consists of mathematical facts which are so delicately balanced between being true or false that we're never going to know, and so you might as well toss a coin. You can't do better than tossing a coin. Which means the chance is half you're going to get it right if you toss the coin and half you'll get it wrong, and you can't really do better than that.

So how do I find this complete lack of structure in an area of pure mathematics? Let me try to give you a quick summary. For those of you who may have heard about it, this is what I like to call Ω, it's a real number, the halting probability.

Omega Number
"Halting Probability"

And some people are nice enough to call this "Chaitin's number." I call it
Ω. So let me try to give you an idea of how you get to this number. By the
way, to show you how much interest there is in Ω, let me mention that this
month there is a very nice article on Ω numbers by Jean-Paul Delahaye in
the French popular science magazine *Pour la Science,* it's in the May 2002
issue.

Well, following Vladimir Tasić, *Mathematics and the Roots of Postmodern
Thought,* the way you explain how to get to this number that shows that some
mathematical facts are true for no reason, they're only true by accident, is you
start with an idea published by Émile Borel in 1927, of using one real number
to answer all possible yes/no questions, not just mathematical questions, all
possible yes/no questions in English—and in Borel's case it was questions in
French. How do you do it?

Well, the idea is you write a list of all possible questions. You make a list
of all possible questions, in English, or in French. A first, a second, a third,
a fourth, a fifth:

Question # 1

Question # 2

Question # 3

Question # 4

Question # 5

The general idea is you order questions say by size, and within questions of
the same size, in some arbitrary alphabetical order. You number all possible
questions.

And then you define a real number, Borel's number, it's defined like this:

Borel's Number
$$.d_1 d_2 d_3 d_4 d_5$$
The Nth digit after the decimal point, d_N,
answers the Nth question!!

Well, you may say, most of these questions are going to be garbage probably,
if you take all possible texts from the English alphabet, or French alphabet.
Yes, but a digit has ten possibilities, so you can let 1 mean the answer is
yes, 2 mean the answer is no, and 3 mean it's not a valid yes/no question in

English, because it's not valid English, or it is valid English, but it's not a question, or it is a valid question, but it's not a yes/no question, for example, it asks for your opinion. There are various ways to deal with all of this.

So you can do all this with one real number—and a real number is a number that's measured with infinite precision, with an infinite number of digits d_N after the decimal point—you can give the answers to all yes/no questions! And these will be questions about history, questions about philosophy, questions about mathematics, questions about physics.

It can do this because there's an awful lot you can put into a real number. It has an infinite amount of information, because it has an infinite number of digits. So this is a way to say that real numbers are very **un**real, right? So let's start with this very unreal number that answers all yes/no questions, and I'll get to my Ω number in a few steps.

The next step is to make it only answer questions about Turing's halting problem. So what's Turing's halting problem? Well, the halting problem is a famous question that Turing considered in 1936. It's about as famous as Gödel's 1931 work, but it's different.

Turing's Halting Problem 1936
[1931 Gödel]

And what Turing showed is that there are limits to mathematical reasoning, but he did it very differently from Gödel, he found something concrete. He doesn't say "this statement is unprovable" like Gödel, he found something concrete that mathematical reasoning can't do: it can't settle in advance whether a computer program will ever halt. This is the halting problem, and it's in a wonderful paper, it's the beginning of theoretical computer science, and it was done before there were computers. And this is the Turing who then went on and did important things in cryptography during the Second World War, and built computers after the war. Turing was a Jack of all trades.

So how do you prove Turing's result that there's no algorithm to decide if a computer program—a self-contained computer program—will ever halt? (Actually the problem is to decide that it will **never** halt.) Well, that's not hard to do, in many different ways, and I sketched a proof before, when I was talking about proving that programs are elegant.

So let's take Borel's real number, and let's change it so that it only answers instances of the halting problem. So you just find a way of numbering all possible computer programs, you pick some fixed language, and you number all programs somehow: first program, second program, third program, you make a list of all possible computer programs in your mind, it's a mental fantasy.

Computer Program # 1
Computer Program # 2
Computer Program # 3
Computer Program # 4
Computer Program # 5

And then what you do is you define a real number whose Nth digit—well, let's make it binary now instead of decimal—whose Nth bit tells us if the Nth computer program ever halts.

Turing's Number
$$.b_1 b_2 b_3 b_4 b_5$$
The Nth bit after the binary point, b_N,
tells us if the Nth computer program ever halts.

So we've already economized a little, we've gone from a decimal number to a binary number. This number is between zero and one, and so is Borel's number, there's no integer part to this real number. It's all in the fractional part. You have an infinite number of digits or bits after the decimal point or the binary point. In the previous number, Borel's original one, the Nth digit answers the Nth yes/no question in French. And here the Nth bit of this new number, Turing's number, will be 0 if the Nth computer program never halts, and it'll be 1 if the Nth computer program does eventually halt.

So this one number would answer all instances of Turing's halting problem. And this number is uncomputable, Turing showed that in his 1936 paper. There's no way to calculate this number, it's an uncomputable real number, because the halting problem is unsolvable. This is shown by Turing in his paper.

So what's the next step? This still doesn't quite get you to randomness. This number gets you to uncomputability. But it turns out this number, Turing's number, is redundant. Why is it redundant?

Redundant

Well, the answer is that there's a lot of repeated information in the bits of this number. We can actually compress it more, we don't have complete randomness yet. Why is there a lot of redundancy? Why is there a lot of repeated information in the bits of this number? Well, because different cases of the halting problem are connected. These bits b_N are not independent of each other. Why?

Well, let's say you have K instances of the halting problem. That is to say, somebody gives you K computer programs and asks you to determine in each case, does it halt or not.

K instances of the halting problem?

Is this K bits of mathematical information? K instances of the halting problem will give us K bits of Turing's number. Are these K bits independent pieces of information? Well, the answer is no, they never are. Why not? Because you don't really need to know K yes/no answers, it's not really K full bits of information. There's a lot less information. It can be compressed. Why?

Well, the answer is very simple. If you have to ask God or an oracle that answers yes/no questions, you don't really need to ask K questions to the oracle, you don't need to bother God that much! You really only need to know what? Well, it's sufficient to know **how many of the programs halt.**

And this is going to be a number between zero and K, a number that's between zero and K.

$$0 \leq \# \text{ that halt} \leq K$$

And if you write this number in binary it's really only about $\log_2 K$ bits.

$$\# \text{ that halt} = \log_2 K \text{ bits}$$

If you know how many of these K programs halt, then what you do is you just start running them all in parallel until you find that precisely that number of programs have halted, and at that point you can stop, because you know the other ones will never halt. And knowing how many of them halt is a lot less than K bits of information, it's really only about $\log_2 K$ bits, it's the number of bits you need to be able to express a number between zero and K in binary, you see.

So different instances of the halting problem are never independent, there's a lot of redundant information, and Turing's number has a lot of redundancy. But essentially just by using this idea of telling how many of them halt, you can squeeze out all the redundancy. You know, the way to get to randomness is to remove redundancy! You distill it, you concentrate it, you crystallize it. So what you do is essentially you just take advantage of this observation—it's a little more sophisticated than that—and what you get is my halting probability.

So let me write down an expression for it. It's defined like this:

Omega Number

$$\Omega = \sum_{p \text{ halts}} 2^{-|p|}$$

$|p|$ = size in bits of program p

$$0 < \Omega < 1$$

Then write Ω in binary!

So this is how you get randomness, this is how you show that there are facts that are true for no reason in pure math. You define this number Ω, and to explain this I would take a long time and I don't have it, so this is just a tease!

For more information you can go to my books. I actually have **four** small books published by Springer-Verlag on this subject: *The Limits of Mathematics, The Unknowable, Exploring Randomness* and *Conversations with a Mathematician.* These books come with LISP software and a Java applet LISP interpreter that you can get at my website.

So you define this Ω number to be what? You pick a computer programming language, and you look at all programs p that halt, p is a program, and you sum over all programs p that halt. And what do you sum? Well, if the program p is K bits long, it contributes $1/2^K$, one over two to the K, to this halting probability.

In other words, each K-bit program has probability $1/2^K$, and you'll immediately notice that there are two to the thousand thousand-bit programs, so probably this sum will diverge and give infinity, if you're not careful. And the answer is yes, you're right if you worry about that. So you have to be careful to do things right, and the basic idea is that no extension of a valid program is a valid program. And if you stipulate that the programming language is like that, that its programs are "self-delimiting," then this sum is in fact between zero and one and everything works. Okay?

Anyway, I don't want to go into the details because I don't have time. So if you do everything right, this sum

$$\sum_{p \text{ halts}} 2^{-|p|}$$

actually converges to a number between zero and one which is the halting probability Ω. This is the probability that a program, each bit of which is generated by an independent toss of a fair coin, eventually halts. And it's a way of summarizing all instances of the halting problem in one real number and doing it so cleverly that there's no redundancy.

So if you take this number and then you write it in binary, this halting probability, it turns out that those bits of this number written in binary, these are independent, irreducible mathematical facts, there's absolutely no structure. Even though there's a simple mathematical definition of Ω, those bits, if you could see them, could not be distinguished from independent tosses of a fair coin. There is no mathematical structure that you would ever be able to detect with a computer, there's no algorithmic pattern, there's no structure that you can capture with mathematical proofs—even though Ω has a simple mathematical definition. It's incompressible, irreducible mathematical information. And the reason is, because if you knew the first N bits of this number Ω, it would solve the halting problem for all programs up to N bits in size, it would enable you to answer the halting problem for all programs p up to N bits in size. That's how you prove that this Ω number is random in the sense I explained before of being algorithmically incompressible information.

And that means that not only you can't compress it into a smaller algorithm, you can't compress it into fewer bits of axioms. So if you wanted to be able to determine K bits of Ω, you'd need K bits of axioms to be able to prove what K bits of this number are. It has—its bits have—no structure or pattern that we are capable of seeing.

However, you **can** prove all kinds of nice mathematical theorems about this Ω number. Even though it's a specific real number, it really mimics independent tosses of a fair coin. So for example you can prove that 0's and 1's happen in the limit exactly fifty percent of the time, each of them. You can prove all kinds of **statistical** properties, but you can't determine **individual** bits!

So this is the strongest version I can come up with of an incompleteness result...

Actually, in spite of this, Cristian Calude, Michael Dinneen and Chi-Kou Shu at the University of Auckland have just succeeded in calculating the first 64 bits of a particular Ω number. The halting probability Ω actually depends on the choice of computer or programming language that you write programs in, and they picked a fairly natural one, and were able to decide which programs less than 85 bits in size halt, and from this to get the first 64 bits of this particular halting probability.

This work by Calude *et alia* is reported on page 27 of the 6 April 2002 issue of the British science weekly *New Scientist,* and it's also described in Delahaye's article in the May 2002 issue of the French monthly *Pour la Science,* and it'll be included in the second edition of Calude's book on *Information and Randomness,* which will be out later this year.

But this doesn't contradict my results, because all I actually show is that an N-bit formal axiomatic theory can't enable you to determine substantially more than N bits of the halting probability. And by N-bit axiomatic theory I mean one for which there is an N-bit program for running through all possible proofs and generating all the theorems. So you might in fact be able to get some initial bits of Ω.

Now, what would Hilbert, Gödel and Turing think about all of this?!

I don't know, but I'll tell you what I think it means, it means that math is different from physics, but it's not that different. This is called the quasi-empirical view of mathematics, and Tymoczko has collected a bunch of interesting papers on this subject, in his book on *New Directions in the Philosophy of Mathematics.* This is also connected with what's called experimental mathematics, a leading proponent of which is Jonathan Borwein, and there's a book announced called *Mathematics by Experiment* by Borwein and Bailey that's going to be about this. The general idea is that proofs are fine, but if you can't find a proof, computational evidence can be useful, too.

Now I'd like to tell you about some questions that I don't know how to answer, but that I think are connected with this stuff that I've been talking about. So let me mention some questions I **don't** know how to answer. They're not easy questions.

Well, one question is positive results on mathematics:

Positive Results
Where do new mathematical concepts come from?

I mean, Gödel's work, Turing's work and my work are negative in a way, they're incompleteness results, but on the other hand, they're positive, because in each case you introduce a new concept: incompleteness, uncomputability and algorithmic randomness. So in a sense they're examples that mathematics goes forward by introducing new concepts! So how about an optimistic theory instead of negative results about the limits of mathematical reasoning? In fact, these negative metamathematical results are taking place in a century which is a tremendous, spectacular success for mathematics, mathematics is advancing by leaps and bounds. So there's no reason for pessimism. So what we need is a more realistic theory that gives us a better idea of **why** mathematics is doing so splendidly, which it is. But I'd like to have some theoretical understanding of this, not just anecdotal evidence, like the book about the Wiles proof of Fermat's result.[1]

So this is one thing that I don't know how to do and I hope somebody will do.

Another thing which I think is connected, isn't where new mathematical ideas come from, it's where do new biological organisms come from. I want a theory of evolution, biological evolution.[2]

Biological Evolution
Where do new biological ideas come from?

You see, in a way biological organisms are ideas, or genes are ideas. And good ideas get reused. You know, it's programming, in a way, biology.

Another question isn't theoretical evolutionary biology—which doesn't exist, but that is what I'd like to see—another question is where do new ideas come from, not just in math! **Our** new ideas. How does the brain work? How does the mind work? Where do new ideas come from? So to answer that, you need to solve the problem of AI or how the brain works!

AI/Brain/Mind
Where do new ideas come from?

In a sense, where new mathematical concepts come from is related to this, and so is the question of the origin of new biological ideas, new genes, new ideas for building organisms—and the ideas keep getting reused. That's how biology seems to work. Nature is a cobbler!—So I think these problems are connected, and I hope they have something to do with the ideas I mentioned, my ideas, but perhaps not in the form that I've presented them here.

So I don't know how to answer these questions, but maybe some of you will be able to answer them. I hope so! The future is yours, do great things!

---

[1]Simon Singh, *Fermat's Enigma;* see also the musical *Fermat's Last Tango.*

[2]In Chapter 12 of *A New Kind of Science,* Stephen Wolfram says that he thinks there is nothing to it, that you get life right away, we're just universal Turing machines, but I think there's more to it than that.

References

1. J. Borwein, D. Bailey, *Mathematics by Experiment*, A. K. Peters, to appear.

2. C. Calude, *Information and Randomness*, Springer-Verlag, 2002.

3. G. J. Chaitin, "Information-theoretic computational complexity," *IEEE Information Theory Transactions*, 1974, pp. 10–15.

4. G. J. Chaitin, "Randomness and mathematical proof," "Randomness in arithmetic," *Scientific American*, May 1975, July 1988, pp. 47–52, 80–85.

5. G. J. Chaitin, *The Limits of Mathematics, The Unknowable, Exploring Randomness, Conversations with a Mathematician*, Springer-Verlag, 1998, 1999, 2001, 2002.

6. M. Chown, "Smash and grab," *New Scientist*, 6 April 2002, pp. 24–28.

7. J.-P. Delahaye, "Les nombres oméga," *Pour la Science*, May 2002, pp. 98–103.

8. G. Polya, *How to Solve It*, Princeton University Press, 1988.

9. J. Rosenblum, J. S. Lessner, *Fermat's Last Tango*, Original Cast Records OC-6010, 2001.

10. S. Singh, *Fermat's Enigma*, Walker and Co., 1997.

11. V. Tasić, *Mathematics and the Roots of Postmodern Thought*, Oxford University Press, 2001.

12. T. Tymoczko, *New Directions in the Philosophy of Mathematics*, Princeton University Press, 1998.

13. D. Wells, *The Penguin Dictionary of Curious and Interesting Numbers*, Penguin Books, 1986.

14. S. Wolfram, *A New Kind of Science*, Wolfram Media, 2002.

Two philosophical applications of algorithmic information theory

*Two philosophical applications of the concept of program-size complexity are discussed. First, we consider the light program-size complexity sheds on whether mathematics is invented or discovered, i.e., is empirical or is **a priori**. Second, we propose that the notion of algorithmic independence sheds light on the question of being and how the world of our experience can be partitioned into separate entities.*

1. Introduction. Why is program size of philosophical interest?

The cover of the January 2003 issue of *La Recherche* asks this dramatic question:

> Dieu est-il un ordinateur? [Is God a computer?]

The long cover story [1] is a reaction to Stephen Wolfram's controversial book *A New Kind of Science* [2]. The first half of the article points out Wolfram's predecessors, and the second half criticizes Wolfram.

The second half of the article begins (p. 38) with these words:

> Il [Wolfram] n'avance aucune raison sérieuse de penser que les complexités de la nature puissent être générées par des règles énonçables sous forme de programmes informatiques simples.

189

The reason for thinking that a simple program might describe the world is, basically, just Plato's postulate that the universe is rationally comprehensible (*Timaeus*). A sharper statement of this principle is in Leibniz's *Discours de métaphysique* [3], section **VI**. Here is Leibniz's original French (1686):

> Mais Dieu a choisi celuy qui est le plus parfait, c'est à dire celuy qui est en même temps le plus simple en hypotheses et le plus riche en phenomenes, comme pourroit estre une ligne de Geometrie dont la construction seroit aisée et les proprietés et effects seroient fort admirables et d'une grande étendue.

For an English translation of this, see [4].

And Hermann Weyl [5] discovered that in *Discours de métaphysique* Leibniz also states that a physical law has no explicative power if it is as complicated as the body of data it was invented to explain.[1]

This is where algorithmic information theory (AIT) comes in. AIT posits that a theory that explains X is a computer program for calculating X, that therefore must be smaller, much smaller, than the size in bits of the data X that it explains. AIT makes a decisive contribution to philosophy by providing a mathematical theory of complexity. AIT defines the *complexity* or *algorithmic information content* of X to be the size in bits $H(X)$ of the smallest computer program for calculating X. $H(X)$ is also the complexity of the most elegant (the simplest) theory for X.

In this article we discuss some other philosophical applications of AIT.

For those with absolutely no background in philosophy, let me recommend two excellent introductions, Magee [6] and Brown [7]. For introductions to AIT, see Chaitin [8, 9]. For another discussion of the philosophical implications of AIT, see Chaitin [10].

2. Is mathematics empirical or is it a priori?

2.1. Einstein: Math is empirical

Einstein was a physicist and he believed that math is invented, not discovered. His sharpest statement on this is his declaration that "the series of

---

[1]See the Leibniz quote in Section 2.3 below.

integers is obviously an invention of the human mind, a self-created tool which simplifies the ordering of certain sensory experiences."

Here is more of the context:

> In the evolution of philosophic thought through the centuries the following question has played a major rôle: What knowledge is pure thought able to supply independently of sense perception? Is there any such knowledge?... I am convinced that... the concepts which arise in our thought and in our linguistic expressions are all... the free creations of thought which can not inductively be gained from sense-experiences... **Thus, for example, the series of integers is obviously an invention of the human mind, a self-created tool which simplifies the ordering of certain sensory experiences.**[2]

The source is Einstein's essay "Remarks on Bertrand Russell's theory of knowledge." It was published in 1944 in the volume [11] on *The Philosophy of Bertrand Russell* edited by Paul Arthur Schilpp, and it was reprinted in 1954 in Einstein's *Ideas and Opinions* [12].

And in his *Autobiographical Notes* [13] Einstein repeats the main point of his Bertrand Russell essay, in a paragraph on Hume and Kant in which he states that "all concepts, even those closest to experience, are from the point of view of logic freely chosen posits." Here is the bulk of this paragraph:

> Hume saw clearly that certain concepts, as for example that of causality, cannot be deduced from the material of experience by logical methods. Kant, thoroughly convinced of the indispensability of certain concepts, took them... to be the necessary premises of any kind of thinking and distinguished them from concepts of empirical origin. I am convinced, however, that this distinction is erroneous or, at any rate, that it does not do justice to the problem in a natural way. **All concepts, even those closest to experience, are from the point of view of logic freely chosen posits...**

---

[2][The boldface emphasis in this and future quotations is mine, not the author's.]

2.2. Gödel: Math is a priori

On the other hand, Gödel was a Platonist and believed that math is a priori. He makes his position blindingly clear in the introduction to an unpublished lecture Gödel *1961/?, "The modern development of the foundations of mathematics in the light of philosophy," *Collected Works* [14], vol. 3:[3]

> I would like to attempt here to describe, in terms of philosophical concepts, the development of foundational research in mathematics..., and to fit it into a general schema of possible philosophical world-views [Weltanschauungen]... I believe that the most fruitful principle for gaining an overall view of the possible world-views will be to divide them up according to the degree and the manner of their affinity to or, respectively, turning away from metaphysics (or religion). In this way we immediately obtain a division into two groups: skepticism, materialism and positivism stand on one side, spiritualism, idealism and theology on the other... Thus one would, for example, say that apriorism belongs in principle on the right and empiricism on the left side... Now it is a familiar fact, even a platitude, that the development of philosophy since the Renaissance has by and large gone from right to left... It would truly be a miracle if this (I would like to say rabid) development had not also begun to make itself felt in the conception of mathematics. Actually, **mathematics, by its nature as an a priori science**, always has, in and of itself, an inclination toward the right, and, for this reason, **has long withstood the spirit of the time** [Zeitgeist] that has ruled since the Renaissance; i.e., the empiricist theory of mathematics, such as the one set forth by Mill, did not find much support... Finally, however, around the turn of the century, its hour struck: in particular, it was the antinomies of set theory, contradictions that allegedly appeared within mathematics, whose significance was exaggerated by skeptics and empiricists and which were employed as a pretext for the leftward upheaval...

---

[3]The numbering scheme used in Gödel's *Collected Works* begins with an * for unpublished papers, followed by the year of publication, or the first/last year that Gödel worked on an unpublished paper.

Nevertheless, the Platonist Gödel makes some remarkably strong statements in favor of adding to mathematics axioms which are not self-evident and which are only justified pragmatically. What arguments does he present in support of these heretical views?

First let's take a look at his discussion of whether Cantor's continuum hypothesis could be established using a new axiom [Gödel 1947, "What is Cantor's continuum problem?", *Collected Works,* vol. 2]:

> ... even **disregarding the intrinsic necessity of some new axiom**, and even in case it has no intrinsic necessity at all, **a probable decision about its truth is possible also in another way, namely, inductively** by studying its "success." Success here means fruitfulness in consequences, in particular in "verifiable" consequences, i.e., consequences demonstrable without the new axiom, whose proofs with the help of the new axiom, however, are considerably simpler and easier to discover, and make it possible to contract into one proof many different proofs. The axioms for the system of real numbers, rejected by intuitionists, have in this sense been verified to some extent, owing to the fact that analytical number theory frequently allows one to prove number-theoretical theorems which, in a more cumbersome way, can subsequently be verified by elementary methods. A much higher degree of verification than that, however, is conceivable. **There might exist axioms** so abundant in their verifiable consequences, shedding so much light upon a whole field, and yielding such powerful methods for solving problems (and even solving them constructively, as far as that is possible) **that**, no matter whether or not they are intrinsically necessary, they **would have to be accepted at least in the same sense as any well-established physical theory**.

Later in the same paper Gödel restates this:

> It was pointed out earlier... that, **besides mathematical intuition, there exists another** (though only probable) **criterion of the truth of mathematical axioms, namely** their **fruitfulness** in mathematics and, one may add, possibly also in

physics... The simplest case of an application of the criterion under discussion arises when some... axiom has number-theoretical consequences verifiable by computation up to any given integer.

And here is an excerpt from Gödel's contribution [Gödel 1944, "Russell's mathematical logic," *Collected Works,* vol. 2] to the same Bertrand Russell festschrift volume [11] that was quoted above:

> The analogy between mathematics and a natural science is enlarged upon by Russell also in another respect... **axioms need not be evident in themselves, but rather their justification lies (exactly as in physics) in the fact that they make it possible for these "sense perceptions" to be deduced...** I think that... this view has been largely justified by subsequent developments, and it is to be expected that it will be still more so in the future. It has turned out that the solution of certain arithmetical problems requires the use of assumptions essentially transcending arithmetic... Furthermore it seems likely that for deciding certain questions of abstract set theory and even for certain related questions of the theory of real numbers new axioms based on some hitherto unknown idea will be necessary. Perhaps also the apparently insurmountable difficulties which some other mathematical problems have been presenting for many years are due to the fact that the necessary axioms have not yet been found. Of course, under these circumstances mathematics may lose a good deal of its "absolute certainty;" but, under the influence of the modern criticism of the foundations, this has already happened to a large extent...

Finally, take a look at this excerpt from Gödel *1951, "Some basic theorems on the foundations," *Collected Works,* vol. 3, an unpublished essay by Gödel:

> I wish to point out that one may conjecture the truth of a universal proposition (for example, that I shall be able to verify a certain property for *any* integer given to me) and at the same time conjecture that no general proof for this fact exists. It is easy to imagine situations in which both these conjectures would

be very well founded. For the first half of it, this would, for example, be the case if the proposition in question were some equation $F(n) = G(n)$ of two number-theoretical functions which could be verified up to *very* great numbers n.[4] Moreover, exactly as in the natural sciences, this *inductio per enumerationem simplicem* is by no means the only inductive method conceivable in mathematics. I admit that every mathematician has an inborn abhorrence to giving more than heuristic significance to such inductive arguments. I think, however, that this is due to the very prejudice that mathematical objects somehow have no real existence. **If mathematics describes an objective world just like physics, there is no reason why inductive methods should not be applied in mathematics just the same as in physics.** The fact is that in mathematics we still have the same attitude today that in former times one had toward all science, namely, we try to derive everything by cogent proofs from the definitions (that is, in ontological terminology, from the essences of things). Perhaps this method, if it claims monopoly, is as wrong in mathematics as it was in physics.

So Gödel the Platonist has nevertheless managed to arrive, at least partially, at what I would characterize, following Tymoczko [16], as a pseudo-empirical or a quasi-empirical position!

2.3. AIT: Math is quasi-empirical

What does algorithmic information theory have to contribute to this discussion? Well, I believe that AIT also supports a quasi-empirical view of mathematics. And I believe that it provides further justification for Gödel's belief that we should be willing to add new axioms.

Why do I say this?

As I have argued on many occasions, AIT, by measuring the complexity (algorithmic information content) of axioms and showing that Gödel incompleteness is natural and ubiquitous, deepens the arguments that forced Gödel,

---

[4]Such a verification of an *equality* (not an inequality) between two number-theoretical **functions of not too complicated or artificial structure** would certainly give a great probability to their complete equality, although its numerical value could not be estimated in the present state of science. However, it is easy to give examples of general propositions about integers where the probability can be estimated even now...

in spite of himself, in spite of his deepest instincts about the nature of mathematics, to believe in inductive mathematics. And if one considers the use of induction rather than deduction to establish mathematical facts, some kind of notion of *complexity* must necessarily be involved. For as Leibniz stated in 1686, a theory is only convincing to the extent that it is substantially *simpler* than the facts it attempts to explain:

> ... non seulement rien n'arrive dans le monde, qui soit absolument irregulier, mais on ne sçauroit mêmes rien feindre de tel. Car supposons par exemple que quelcun fasse quantité de points sur le papier à tout hazard, comme font ceux qui exercent l'art ridicule de la Geomance, je dis qu'il est possible de trouver une ligne geometrique dont la motion soit constante et uniforme suivant une certaine regle, en sorte que cette ligne passe par tous ces points... Mais **quand une regle est fort composée, ce qui luy est conforme, passe pour irrégulier**. Ainsi on peut dire que de quelque maniere que Dieu auroit créé le monde, il auroit tousjours esté regulier et dans un certain ordre general. Mais Dieu a choisi celuy qui est le plus parfait, c'est à dire celuy qui est en même temps **le plus simple en hypotheses et le plus riche en phenomenes**... [*Discours de métaphysique*, **VI**]

In fact Gödel himself, in considering inductive rather than deductive mathematical proofs, began to make some tentative initial attempts to formulate and utilize notions of complexity. (I'll tell you more about this in a moment.) And it is here that AIT makes its decisive contribution to philosophy, by providing a highly-developed and elegant mathematical theory of complexity. How does AIT do this? It does this by considering the size of the smallest computer program required to calculate a given object X, which may also be considered to be the most elegant theory that explains X.

Where does Gödel begin to think about complexity? He does so in two footnotes in vol. 3 of his *Collected Works*. The first of these is a footnote to Gödel *1951. This footnote begins "Such a verification..." and it was reproduced, in part, in Section 2.2 above. And here is the relevant portion of the second, the more interesting, of these two footnotes:

> ... Moreover, if every number-theoretical question of Goldbach type... is decidable by a mathematical proof, there *must* exist an

infinite set of independent evident axioms, i.e., a set m of evident axioms which are not derivable from *any* finite set of axioms (no matter whether or not the latter axioms belong to m and whether or not they are evident). Even if solutions are desired only for all those problems of Goldbach type which are simple enough to be formulated in a few pages, **there must exist** a great number of evident axioms or evident **axioms of great complication, in contradistinction to the few simple axioms upon which all of present day mathematics is built.** (It can be proved that, in order to solve all problems of Goldbach type of a certain degree of complication k, one needs a system of axioms whose degree of complication, up to a minor correction, is $\geq k$.)[5]

This is taken from Gödel *1953/9–III, one of the versions of his unfinished paper "Is mathematics syntax of language?" that was intended for, but was finally not included, in Schilpp's Carnap festschrift in the same series as the Bertrand Russell festschrift [11].

Unfortunately these tantalizing glimpses are, as far as I'm aware, all that we know about Gödel's thoughts on complexity. Perhaps volumes 4 and 5, the two final volumes of Gödel's *Collected Works,* which contain Gödel's correspondence with other mathematicians, and which will soon be available, will shed further light on this.

Now let me turn to a completely different—but I believe equally fundamental—application of AIT.

3. How can we partition the world into distinct entities?

For many years I have asked myself, "What is a living being? How can we define this mathematically?!" I still don't know the answer! But at least

---

[5][This is reminiscent of the theorem in AIT that p_k = (the program of size $\leq k$ bits that takes longest to halt) is the simplest possible "axiom" from which one can solve the halting problem for all programs of size $\leq k$. Furthermore, p_k's size and complexity both differ from k by at most a fixed number of bits: $|p_k| = k + O(1)$ and $H(p_k) = k + O(1)$.

Actually, in order to solve the halting problem for all programs of size $\leq k$, in addition to p_k one needs to know $k - |p_k|$, which is how much p_k's size differs from k. This fixed amount of additional information is required in order to be able to determine k from p_k.]

Of course, these observations are just the beginning. A great deal more work is needed to develop this point of view...

For a technical discussion of algorithmic independence and the associated notion of *mutual algorithmic information* defined as follows

$$H(X : Y) \equiv H(X) + H(Y) - H(X, Y),$$

see my book Chaitin [17].

4. Conclusion and future prospects

Let's return to our starting point, to the cover of the January 2003 issue of *La Recherche*. Is God a computer, as Wolfram and some others think, or is God, as Plato and Pythagoras affirm, a mathematician?

And, an important part of this question, **is the physical universe discrete**, the way computers prefer, **not continuous**, the way it seems to be in classical Newtonian/Maxwellian physics? Speaking personally, I like the discrete, not the continuous. And my theory, AIT, deals with discrete, digital information, bits, not with continuous quantities. But the physical universe is of course free to do as it likes!

Hopefully pure thought will not be called upon to resolve this. Indeed, I believe that it is incapable of doing so; Nature will have to tell us. Perhaps someday an *experimentum crucis* will provide a definitive answer. In fact, for a hundred years quantum physics has been pointing insistently in the direction of discreteness.[6]

References

[1] O. Postel-Vinay, "L'Univers est-il un calculateur?" [Is the universe a calculator?], *La Recherche*, no. 360, January 2003, pp. 33–44.

[2] S. Wolfram, *A New Kind of Science*, Wolfram Media, 2002.

[3] Leibniz, *Discours de métaphysique*, Gallimard, 1995.

[4] G. W. Leibniz, *Philosophical Essays*, Hackett, 1989.

---

[6]Discreteness in physics actually began even earlier, with atoms. And then, my colleague John Smolin points out, when Boltzmann introduced coarse-graining in statistical mechanics.

[5] H. Weyl, *The Open World,* Yale University Press, 1932, Ox Bow Press, 1989.

[6] B. Magee, *Confessions of a Philosopher,* Modern Library, 1999.

[7] J. R. Brown, *Philosophy of Mathematics,* Routledge, 1999.

[8] G. J. Chaitin, "Paradoxes of randomness," *Complexity,* vol. 7, no. 5, pp. 14–21, 2002.

[9] G. J. Chaitin, "Meta-mathematics and the foundations of mathematics," *Bulletin EATCS,* vol. 77, pp. 167–179, 2002.

[10] G. J. Chaitin, "On the intelligibility of the universe and the notions of simplicity, complexity and irreducibility," `http://arxiv.org/math.HO/0210035`, 2002.

[11] P. A. Schilpp, *The Philosophy of Bertrand Russell,* Open Court, 1944.

[12] A. Einstein, *Ideas and Opinions,* Crown, 1954, Modern Library, 1994.

[13] A. Einstein, *Autobiographical Notes,* Open Court, 1979.

[14] K. Gödel, *Collected Works,* vols. 1–5, Oxford University Press, 1986–2003.

[15] *Kurt Gödel: Wahrheit & Beweisbarkeit* [truth and provability], vols. 1–2, öbv & hpt, 2002.

[16] T. Tymoczko, *New Directions in the Philosophy of Mathematics,* Princeton University Press, 1998.

[17] G. J. Chaitin, *Exploring Randomness,* Springer-Verlag, 2001.

On the intelligibility of the universe and the notions of simplicity, complexity and irreducibility

We discuss views about whether the universe can be rationally comprehended, starting with Plato, then Leibniz, and then the views of some distinguished scientists of the previous century. Based on this, we defend the thesis that comprehension is compression, i.e., explaining many facts using few theoretical assumptions, and that a theory may be viewed as a computer program for calculating observations. This provides motivation for defining the complexity of something to be the size of the simplest theory for it, in other words, the size of the smallest program for calculating it. This is the central idea of algorithmic information theory (AIT), a field of theoretical computer science. Using the mathematical concept of program-size complexity, we exhibit irreducible mathematical facts, mathematical facts that cannot be demonstrated using any mathematical theory simpler than they are. It follows that the world of mathematical ideas has infinite complexity and is therefore not fully comprehensible, at least not in a static fashion. Whether the physical world has finite or infinite complexity remains to be seen. Current science believes that the world contains randomness, and is therefore also infinitely complex, but a deterministic universe that simulates randomness via pseudo-randomness is also a possibility, at least according to recent highly speculative work of S. Wolfram. [Written for a meeting of the German Philosophical Society, Bonn, September 2002.]

"Nature uses only the longest threads to weave her patterns, so that each small piece of her fabric reveals the organization of the entire tapestry."

—Feynman, *The Character of Physical Law,* 1965, at the very end of Chapter 1, "The Law of Gravitation".[1]

"The most incomprehensible thing about the universe is that it is comprehensible."

—Attributed to Einstein. The original source, where the wording is somewhat different, is Einstein, "Physics and Reality", 1936, reprinted in Einstein, *Ideas and Opinions,* 1954.[2]

It's a great pleasure for me to speak at this meeting of the German Philosophical Society. Perhaps it's not generally known that at the end of his life my predecessor Kurt Gödel was obsessed with Leibniz.[3] Writing this paper was for me a voyage of discovery—of the depth of Leibniz's thought! Leibniz's power as a philosopher is informed by his genius as a mathematician; as I'll explain, some of the key ideas of AIT are clearly visible in embryonic form in his 1686 *Discourse on Metaphysics.*

I Plato's *Timaeus*—The Universe is Intelligible. Origins of the Notion of Simplicity: Simplicity as Symmetry [Brisson, Meyerstein 1991]

"[T]his is the central idea developed in the *Timaeus*: the order established by the demiurge in the universe becomes manifest as the symmetry found at its most fundamental level, a symmetry which makes possible a mathematical description of such a universe."

—Brisson, Meyerstein, *Inventing the Universe,* 1995 (1991 in French). This book discusses the cosmology of Plato's *Timaeus,* modern cosmology and AIT; one of their key insights is to identify symmetry with simplicity.

According to Plato, the world is rationally understandable because it has structure. And the universe has structure, because it is a work of art created

---

[1] An updated version of this chapter would no doubt include a discussion of the infamous astronomical missing mass problem.

[2] Einstein actually wrote "Das ewig Unbegreifliche an der Welt ist ihre Begreiflichkeit". Translated word for word, this is "The eternally incomprehensible about the world is its comprehensibility". But I prefer the version given above, which emphasizes the paradox.

[3] See Menger, *Reminiscences of the Vienna Circle and the Mathematical Colloquium,* 1994.

by a God who is a mathematician. Or, more abstractly, the structure of the world consists of God's thoughts, which are mathematical. The fabric of reality is built out of eternal mathematical truth. [Brisson, Meyerstein, *Inventer l'Univers,* 1991]

Timaeus postulates that simple, symmetrical geometrical forms are the building blocks for the universe: the circle and the regular solids (cube, tetrahedron, icosahedron, dodecahedron, octahedron).

What was the evidence that convinced the ancient Greeks that the world is comprehensible? Partly it was the beauty of mathematics, particularly geometry and number theory, and partly the Pythagorean work on the physics of stringed instruments and musical tones, and in astronomy, the regularities in the motions of the planets and the starry heavens and eclipses. Strangely enough, mineral crystals, whose symmetries magnify enormously quantum-mechanical symmetries that are found at the atomic and molecular level, are never mentioned.

What is our current cosmology?

Since the chaos of everyday existence provides little evidence of simplicity, biology is based on chemistry is based on physics is based on high-energy or particle physics. The attempt to find underlying simplicity and pattern leads reductionist modern science to break things into smaller and smaller components in an effort to find the underlying simple building blocks.

And the modern version of the cosmology of *Timaeus* is the application of symmetries or group theory to understand sub-atomic particles (formerly called elementary particles), for example, Gell-Mann's eightfold way, which predicted new particles. This work classifying the "particle zoo" also resembles Mendeleev's periodic table of the elements that organizes their chemical properties so well.[4]

And modern physicists have also come up with a possible answer to the Einstein quotation at the beginning of this paper. Why do they think that the universe is comprehensible? They invoke the so-called "anthropic principle" [Barrow, Tipler, *The Anthropic Cosmological Principle,* 1986], and declare that we would not be here to ask this question unless the universe had enough order for complicated creatures like us to evolve!

Now let's proceed to the next major step in the evolution of ideas on

---

[4]For more on this, see the essay by Freeman Dyson on "Mathematics in the Physical Sciences" in COSRIMS, *The Mathematical Sciences,* 1969. This is an article of his that was originally published in *Scientific American.*

simplicity and complexity, which is a stronger version of the Platonic creed due to Leibniz.

II What Does it Mean for the Universe to Be Intelligible? Leibniz's Discussion of Simplicity, Complexity and Lawlessness [Weyl 1932]

"As for the simplicity of the ways of God, this holds properly with respect to his means, as opposed to the variety, richness, and abundance, which holds with respect to his ends or effects."

"But, **when a rule is extremely complex, what is in conformity with it passes for irregular.** Thus, one can say, in whatever manner God might have created the world, it would always have been regular and in accordance with a certain general order. But **God has chosen the most perfect world, that is, the one which is at the same time the simplest in hypotheses and the richest in phenomena**, as might be a line in geometry whose construction is easy and whose properties and effects are extremely remarkable and widespread."

—Leibniz, *Discourse on Metaphysics*, 1686, Sections 5–6, from Leibniz, *Philosophical Essays*, edited and translated by Ariew and Garber, 1989, pp. 38–39.

"The assertion that nature is governed by strict laws is devoid of all content if we do not add the statement that it is governed by mathematically simple laws... That **the notion of law becomes empty when an arbitrary complication is permitted** was already pointed out by Leibniz in his *Metaphysical Treatise* [*Discourse on Metaphysics*]. Thus simplicity becomes a working principle in the natural sciences... The astonishing thing is not that there exist natural laws, but that the further the analysis proceeds, the finer the details, the finer the elements to which the phenomena are reduced, the simpler—and not the more complicated, as one would originally expect—the fundamental relations become and the more exactly do they describe the actual occurrences. But this circumstance is apt to weaken the metaphysical power of determinism, since it makes the meaning of natural law depend on the fluctuating distinction between mathematically simple and complicated functions or classes of functions."

—Hermann Weyl, *The Open World, Three Lectures on the Metaphysical Implications of Science*, 1932, pp. 40–42. See a similar discussion on pp. 190–191 of Weyl, *Philosophy of Mathematics and Natural Science*, 1949, Section 23A, "Causality and Law".[5]

"Weyl said, not long ago, that 'the problem of simplicity is of central importance for the epistemology of the natural sciences'. Yet it seems that interest in the problem has lately declined; perhaps because, especially after Weyl's penetrating analysis, there seemed to be so little chance of solving it."

---

[5]This is a remarkable anticipation of my definition of "algorithmic randomness", as a set of observations that only has what Weyl considers to be unacceptable theories, ones that are as complicated as the observations themselves, without any "compression".

—Weyl, *Philosophy of Mathematics and Natural Science,* 1949, p. 155, quoted in Popper, *The Logic of Scientific Discovery,* 1959, Chapter VII, "Simplicity", p. 136.

In his novel *Candide,* Voltaire ridiculed Leibniz, caricaturing Leibniz's subtle views with the memorable phrase **"this is the best of all possible worlds"**. Voltaire also ridiculed the efforts of Maupertius to develop a physics in line with Leibniz's views, one based on a principle of least effort.

Nevertheless versions of least effort play a fundamental role in modern science, starting with Fermat's deduction of the laws for reflection and refraction of light from a principle of least time. This continues with the Lagrangian formulation of mechanics, stating that the actual motion minimizes the integral of the difference between the potential and the kinetic energy. And least effort is even important at the current frontiers, such as in Feynman's path integral formulation of quantum mechanics (electron waves) and quantum electrodynamics (photons, electromagnetic field quanta).[6]

However, all this modern physics refers to versions of least effort, not to ideas, not to information, and not to complexity—which are more closely connected with Plato's original emphasis on symmetry and intellectual simplicity = intelligibility. An analogous situation occurs in theoretical computer science, where work on computational complexity is usually focussed on time, not on the complexity of ideas or information. Work on time complexity is of great practical value, but I believe that the complexity of ideas is of greater conceptual significance. Yet another example of the effort/information divide is the fact that I am interested in the irreducibility of ideas (see Sections V and VI), while Stephen Wolfram (who is discussed later in this section) instead emphasizes time irreducibility, physical systems for which there are no predictive short-cuts and the fastest way to see what they do is just to run them.

Leibniz's doctrine concerns more than "least effort", it also implies that the ideas that produce or govern this world are as beautiful and as simple as possible. In more modern terms, God employed the smallest possible amount of intellectual material to build the world, and the laws of physics are as simple and as beautiful as they can be and allow us, intelligent beings,

---

[6]See the short discussion of minimum principles in Feynman, *The Character of Physical Law,* 1965, Chapter 2, "The Relation of Mathematics to Physics". For more information, see *The Feynman Lectures on Physics,* 1963, Vol. 1, Chapter 26, "Optics: The Principle of Least Time", Vol. 2, Chapter 19, "The Principle of Least Action".

to evolve.[7] The belief in this Leibnizean doctrine lies behind the continuing reductionist efforts of high-energy physics (particle physics) to find the ultimate components of reality. The continuing vitality of this Leibnizean doctrine also lies behind astrophysicist John Barrow's emphasis in his "Theories of Everything" essay on finding the minimal TOE that explains the universe, a TOE that is as simple as possible, with no redundant elements (see Section VII below).

Important point: To say that the fundamental laws of physics must be simple does not at all imply that it is easy or fast to deduce from them how the world works, that it is quick to make predictions from the basic laws. The *apparent complexity* of the world we live in—a phrase that is constantly repeated in Wolfram, *A New Kind of Science,* 2002—then comes from the long deductive path from the basic laws to the level of our experience.[8] So again, I claim that minimum information is more important than minimum time, which is why in Section IV I do not care how long a minimum-size program takes to produce its output, nor how much time it takes to calculate experimental data using a scientific theory.

More on Wolfram: In *A New Kind of Science,* Wolfram reports on his systematic computer search for simple rules with very complicated consequences, very much in the spirit of Leibniz's remarks above. First Wolfram amends the Pythagorean insight that Number rules the universe to assert the primacy of Algorithm, not Number. And those are **discrete** algorithms, it's a digital philosophy![9] Then Wolfram sets out to survey all possible worlds, at least all the simple ones.[10] Along the way he finds a lot of interesting stuff. For example, Wolfram's cellular automata rule 110 is a universal computer, an amazingly simple one, that can carry out **any** computation. *A New Kind of Science* is an attempt to discover the laws of the universe by pure thought, to search systematically for God's building blocks!

The limits of reductionism: In what sense can biology and psychology be reduced to mathematics and physics?! This is indeed the acid test of a reductionist viewpoint! Historical contingency is often invoked here: life as "frozen accidents" (mutations), not something fundamental [Wolfram,

---

[7]This is a kind of "anthropic principle", the attempt to deduce things about the universe from the fact that we are here and able to look at it.

[8]It could also come from the complexity of the initial conditions, or from coin-tossing, i.e., randomness.

[9]That's a term invented by Edward Fredkin, who has worked on related ideas.

[10]That's why his book is so thick!

Gould]. Work on artificial life (Alife) plus advances in robotics are particularly aggressive reductionist attempts. The normal way to "explain" life is evolution by natural selection, ignoring Darwin's own sexual selection and symbiotic/cooperative views of the origin of biological progress—new species—notably espoused by Lynn Margulis ("symbiogenesis"). Other problems with Darwinian gradualism: following the DNA as software paradigm, small changes in DNA software can produce big changes in organisms, and a good way to build this software is by trading useful subroutines (this is called horizontal or lateral DNA transfer).[11] In fact, there is a lack of fossil evidence for many intermediate forms,[12] which is evidence for rapid production of new species (so-called "punctuated equilibrium").

III What do Working Scientists Think about Simplicity and Complexity?

"Science itself, therefore, may be regarded as a minimal problem, consisting of the completest possible presentment of facts with the *least possible expenditure of thought...* Those ideas that hold good throughout the widest domains of research and that supplement the greatest amount of experience, are *the most scientific*."

—Ernst Mach, *The Science of Mechanics,* 1893, Chapter IV, Section IV, "The Economy of Science", reprinted in Newman, *The World of Mathematics,* 1956.

"Furthermore, the attitude that theoretical physics does not explain phenomena, but only classifies and correlates, is today accepted by most theoretical physicists. This means that the criterion of success for such a theory is simply whether it can, by a simple and elegant classifying and correlating scheme, cover very many phenomena, which without this scheme would seem complicated and heterogeneous, and whether the scheme even covers phenomena which were not considered or even not known at the time when the scheme was evolved. (These two latter statements express, of course, the unifying and the predicting power of a theory.)"

—John von Neumann, "The Mathematician", 1947, reprinted in Newman, *The World of Mathematics,* 1956, and in Bródy, Vámos, *The Neumann Compendium,* 1995.

"**These fundamental concepts and postulates, which cannot be further reduced logically, form the essential part of a theory, which reason cannot touch. It is the grand object of all theory to make these irreducible elements as simple and as few in number as possible...** [As] the distance in thought between the fundamental concepts and laws on the one side and, on the other, the conclusions which have to be brought into relation with our experience grows larger and larger, the simpler the

---

[11]This is how bacteria acquire immunity to antibiotics.

[12]Already noted by Darwin.

logical structure becomes—that is to say, the smaller the number of logically independent conceptual elements which are found necessary to support the structure."

—Einstein, "On the Method of Theoretical Physics", 1934, reprinted in Einstein, *Ideas and Opinions,* 1954.

"The aim of science is, on the one hand, a comprehension, as *complete* as possible, of the connection between the sense experiences in their totality, and, on the other hand, the accomplishment of this aim *by the use of a minimum of primary concepts and relations.* (Seeking as far as possible, logical unity in the world picture, i.e., paucity in logical elements.)"

"Physics constitutes a logical system of thought which is in a state of evolution, whose basis cannot be distilled, as it were, from experience by an inductive method, but can only be arrived at by free invention... Evolution is proceeding in the direction of increased simplicity of the logical basis. In order further to approach this goal, we must resign to the fact that the logical basis departs more and more from the facts of experience, and that the path of our thought from the fundamental basis to those derived propositions, which correlate with sense experiences, becomes continually harder and longer."

—Einstein, "Physics and Reality", 1936, reprinted in Einstein, *Ideas and Opinions,* 1954.

"[S]omething general will have to be said... about the points of view from which physical theories may be analyzed critically... The first point of view is obvious: the theory must not contradict empirical facts... The second point of view is not concerned with the relationship to the observations but with the premises of the theory itself, with what may briefly but vaguely be characterized as the 'naturalness' or 'logical simplicity' of the premises (the basic concepts and the relations between these)... We prize a theory more highly if, from the logical standpoint, it does not involve an arbitrary choice among theories that are equivalent and possess analogous structures... I must confess herewith that I cannot at this point, and perhaps not at all, replace these hints by more precise definitions. I believe, however, that a sharper formulation would be possible."

—Einstein, "Autobiographical Notes", originally published in Schilpp, *Albert Einstein, Philosopher-Scientist,* 1949, and reprinted as a separate book in 1979.

"What, then, impels us to devise theory after theory? Why do we devise theories at all? The answer to the latter question is simply: because **we enjoy 'comprehending,' i.e., reducing phenomena by the process of logic to something already known** or (apparently) evident. New theories are first of all necessary when we encounter new facts which cannot be 'explained' by existing theories. But this motivation for setting up new theories is, so to speak, trivial, imposed from without. There is another, more subtle motive of no less importance. This is the striving toward unification and simplification of the premises of the theory as a whole (i.e., Mach's principle of economy, interpreted as a logical principle)."

"There exists a passion for comprehension, just as there exists a passion for music. That passion is rather common in children, but gets lost in most people later on. Without this passion, there would be neither mathematics nor natural science. Time and again the passion for understanding has led to the illusion that man is able to comprehend the objective world rationally, by pure thought, without any empirical foundations—in short, by metaphysics. I believe that every true theorist is a kind of tamed metaphysicist, no matter how pure a 'positivist' he may fancy himself. **The metaphysicist believes that the logically simple is also the real. The tamed metaphysicist believes that not all that is logically simple is embodied in experienced reality, but that the totality of all sensory experience can be 'comprehended' on the basis of a conceptual system built on premises of great simplicity.** The skeptic will say that this is a 'miracle creed.' Admittedly so, but it is a miracle creed which has been borne out to an amazing extent by the development of science."

—Einstein, "On the Generalized Theory of Gravitation", 1950, reprinted in Einstein, *Ideas and Opinions,* 1954.

"One of the most important things in this 'guess—compute consequences—compare with experiment' business is to know when you are right. It is possible to know when you are right way ahead of checking all the consequences. You can recognize truth by its beauty and simplicity. It is always easy when you have made a guess, and done two or three little calculations to make sure that it is not obviously wrong, to know that it is right. **When you get it right, it is obvious that it is right**—at least if you have any experience—**because usually what happens is that more comes out than goes in.** Your guess is, in fact, that something is very simple. If you cannot see immediately that it is wrong, and it is simpler than it was before, then it is right. The inexperienced, and crackpots, and people like that, make guesses that are simple, but you can immediately see that they are wrong, so that does not count. Others, the inexperienced students, make guesses that are very complicated, and it sort of looks as if it is all right, but I know it is not true because the truth always turns out to be simpler than you thought. What we need is imagination, but imagination in a terrible strait-jacket. We have to find a new view of the world that has to agree with everything that is known, but disagree in its predictions somewhere, otherwise it is not interesting. And in that disagreement it must agree with nature..."

—Feynman, *The Character of Physical Law,* 1965, Chapter 7, "Seeking New Laws".

"It is natural that a man should consider the work of his hands or his brain to be useful and important. Therefore nobody will object to an ardent experimentalist boasting of his measurements and rather looking down on the 'paper and ink' physics of his theoretical friend, who on his part is proud of his lofty ideas and despises the dirty fingers of the other. But in recent years this kind of friendly rivalry has changed into something more serious... [A] school of extreme experimentalists... has gone so far as to reject theory altogether... There is also a movement in the opposite direction... claiming that to the mind well trained in mathematics and epistemology the laws of Nature are manifest without appeal to experiment."

"Given the knowledge and the penetrating brain of our mathematician, Maxwell's equations are a result of pure thinking and the toil of experimenters antiquated and superfluous. I need hardly explain to you the fallacy of this standpoint. It lies in the fact that none of the notions used by the mathematicians, such as potential, vector potential, field vectors, Lorentz transformations, quite apart from the principle of action itself, are evident or given *a priori.* Even if an extremely gifted mathematician had constructed them to describe the properties of a possible world, neither he nor anybody else would have had the slightest idea how to apply them to the real world."

"Charles Darwin, my predecessor in my Edinburgh chair, once said something like this: 'The Ordinary Man can see a thing an inch in front of his nose; a few can see things 2 inches distant; if anyone can see it at 3 inches, he is a man of genius.' I have tried to describe to you some of the acts of these 2- or 3-inch men. My admiration of them is not diminished by the consciousness of the fact that they were guided by the experience of the whole human race to the right place into which to poke their noses. **I have also not endeavoured to analyse the idea of beauty or perfection or simplicity of a natural law** which has often guided the correct divination. **I am convinced that such an analysis would lead to nothing; for these ideas are themselves subject to development. We learn something new from every new case**, and I am not inclined to accept final theories about invariable laws of the human mind."

"My advice to those who wish to learn the art of scientific prophecy is not to rely on abstract reason, but to decipher the secret language of Nature from Nature's documents, the facts of experience."

—Max Born, *Experiment and Theory in Physics,* 1943, pp. 1, 8, 34–35, 44.

These eloquent discussions of the role that simplicity and complexity play in scientific discovery by these distinguished 20th century scientists show the importance that they ascribe to these questions.

In my opinion, the fundamental point is this: The belief that the universe is rational, lawful, is of no value if the laws are too complicated for us to comprehend, and is even meaningless if the laws are as complicated as our observations, since the laws are then no simpler than the world they are supposed to explain. As we saw in the previous section, this was emphasized (and attributed to Leibniz) by Hermann Weyl, a fine mathematician and mathematical physicist.

But perhaps we are overemphasizing the role that the notions of simplicity and complexity play in science?

In his beautiful 1943 lecture published as a small book on *Experiment and Theory in Physics,* the theoretical physicist Max Born criticized those who think that we can understand Nature by pure thought, without hints from experiments. In particular, he was referring to now forgotten and rather fanciful theories put forth by Eddington and Milne. Now he might level these criticisms at string theory and at Stephen Wolfram's *A New Kind of Science* [Jacob T. Schwartz, private communication].

Born has a point. Perhaps the universe **is** complicated, not simple! This certainly seems to be the case in biology more than in physics. Then thought alone is insufficient; we need empirical data. But simplicity certainly reflects what we mean by understanding: **understanding is compression.** So perhaps this is more about the human mind than it is about the universe. Perhaps our emphasis on simplicity says more about us than it says about the universe!

Now we'll try to capture some of the essential features of these philosophical ideas in a mathematical theory.

IV A Mathematical Theory of Simplicity, Complexity and Irreducibility: AIT

The basic idea of algorithmic information theory (AIT) is that a scientific theory is a computer program, and the smaller, the more concise the program is, the better the theory!

But the idea is actually much broader than that. **The central idea of algorithmic information theory is reflected in the belief that the following diagrams all have something fundamental in common.** In each case, ask how much information we put in versus how much we get out. And everything is digital, discrete.

Shannon information theory (communications engineering), noiseless coding:

$$\text{encoded message} \to \textbf{Decoder} \to \text{original message}$$

Model of scientific method:

$$\text{scientific theory} \to \textbf{Calculations} \to \text{empirical/experimental data}$$

Algorithmic information theory (AIT), definition of program-size complexity:

$$\text{program} \to \textbf{Computer} \to \text{output}$$

Central dogma of molecular biology:

$$\text{DNA} \to \textbf{Embryogenesis/Development} \to \text{organism}$$

(In this connection, see Küppers, *Information and the Origin of Life,* 1990.) Turing/Post abstract formulation of a Hilbert-style formal axiomatic mathematical theory as a mechanical procedure for systematically deducing all possible consequences from the axioms:

$$\text{axioms} \to \textbf{Deduction} \to \text{theorems}$$

Contemporary physicists' efforts to find a Theory of Everything (TOE):

$$\text{TOE} \rightarrow \textbf{Calculations} \rightarrow \text{Universe}$$

Leibniz, *Discourse on Metaphysics,* 1686:

$$\text{Ideas} \rightarrow \textbf{Mind of God} \rightarrow \text{The World}$$

In each case the left-hand side is smaller, much smaller, than the right-hand side. In each case, the right-hand side can be constructed (re-constructed) mechanically, or systematically, from the left-hand side. And in each case we want to keep the right-hand side fixed while making the left-hand side as small as possible. Once this is accomplished, we can use the size of the left-hand side as a measure of the simplicity or the complexity of the corresponding right-hand side.

Starting with this one simple idea, of looking at the size of computer programs, or at program-size complexity, you can develop a sophisticated, elegant mathematical theory, AIT, as you can see in my four Springer-Verlag volumes listed in the bibliography of this paper.

But, I must confess that AIT makes a large number of **important hidden assumptions!** What are they?

Well, one important hidden assumption of AIT is that the choice of computer or of computer programming language is not too important, that it does not affect program-size complexity too much, in any fundamental way. This is debatable.

Another important tacit assumption: we use the discrete computation approach of Turing 1936, eschewing computations with "real" (infinite-precision) numbers like $\pi = 3.1415926\ldots$ which have an infinite number of digits when written in decimal notation, but which correspond, from a geometrical point of view, to a single point on a line, an elemental notion in continuous, but not in discrete, mathematics. Is the universe **discrete** or **continuous**? Leibniz is famous for his work on continuous mathematics. AIT sides with the discrete, not with the continuous. [Françoise Chaitin-Chatelin, private communication]

Also, in AIT we completely ignore the **time** taken by a computation, concentrating only on the **size** of the program. And the computation run-times may be monstrously large, quite impracticably so, in fact, totally astronomical in size. But trying to take time into account destroys AIT, an elegant, simple theory of complexity, and one which imparts much intuitive understanding. So I think that it is a mistake to try to take time into account when thinking about this kind of complexity.

We've talked about simplicity and complexity, but what about **irreducibility**? Now let's apply AIT to mathematical logic and obtain some limitative metatheorems. However, following Turing 1936 and Post 1944, I'll use the notion of algorithm to deduce limits to formal reasoning, not Gödel's original 1931 approach. I'll take the position that a Hilbert-style mathematical theory, a formal axiomatic theory, is a mechanical procedure for systematically generating all the theorems by running through all possible proofs, systematically deducing all consequences of the axioms.[13] Consider the size in bits of the algorithm for doing this. This is how we measure the simplicity or complexity of the formal axiomatic theory. It's just another instance of program-size complexity!

But at this point, Chaitin-Chatelin insists, I should admit that we are making an extremely embarrassing hidden assumption, which is that you can systematically run through all the proofs. This assumption, which is bundled into my definition of a formal axiomatic theory, means that we are assuming that the language of our theory is static, and that no new concepts can ever emerge. But no human language or field of thought is static![14] And this idea of being able to make a numbered list with all possible proofs was clearly anticipated by Émile Borel in 1927 when he pointed out that there is a real number with the problematical property that its Nth digit after the decimal point gives us the answer to the Nth yes/no question in French.[15]

Yes, I agree, a Hilbert-style formal axiomatic theory is indeed a fantasy, but it is a fantasy that inspired many people, and one that even helped to lead to the creation of modern programming languages. It is a fantasy that it is useful to take seriously long enough for us to show in Section VI that even if you are willing to accept all these tacit assumptions, something else is terribly wrong. Formal axiomatic theories can be criticized from within, as well as from without. And it is far from clear how weakening these tacit assumptions would make it easier to prove the irreducible mathematical truths that are exhibited in Section VI.

---

[13]In a way, this point of view was anticipated by Leibniz with his *lingua characteristica universalis*.

[14]And computer programming languages aren't static either, which can be quite a nuisance.

[15]Borel's work was brought to my attention by Vladimir Tasić in his book *Mathematics and the Roots of Postmodern Thought*, 2001, where he points out that in some ways it anticipates the Ω number that I'll discuss in Section IX. Borel's paper is reprinted in Mancosu, *From Brouwer to Hilbert*, 1998, pp. 296–300.

And the idea of a fixed, static computer programming language in which you write the computer programs whose size you measure is also a fantasy. Real computer programming languages don't stand still, they evolve, and the size of the computer program you need to perform a given task can therefore change. Mathematical models of the world like these are always approximations, "lies that help us to see the truth" (Picasso). Nevertheless, if done properly, they can impart insight and understanding, they can help us to comprehend, they can reveal unexpected connections...

V From Computational Irreducibility to Logical Irreducibility. Examples of Computational Irreducibility: "Elegant" Programs

Our goal in this section and the next is to use AIT to establish the existence of **irreducible mathematical truths**. What are they, and why are they important?

Following Euclid's *Elements,* a mathematical truth is established by reducing it to simpler truths until self-evident truths—"axioms" or "postulates"[16]—are reached. Here we exhibit an extremely large class of mathematical truths that are not at all self-evident but which are **not** consequences of any principles simpler than they are.

Irreducible truths are highly problematical for traditional philosophies of mathematics, but as discussed in Section VIII, they can be accommodated in an emerging "quasi-empirical" school of the foundations of mathematics, which says that physics and mathematics are not that different.

Our path to logical irreducibility starts with computational irreducibility. Let's start by calling a computer program "elegant" if no smaller program in the same language produces exactly the same output. There are lots of elegant programs, at least one for each output. And it doesn't matter how **slow** an elegant program is, all that matters is that it be as **small** as possible.

An elegant program viewed as an object in its own right is computationally irreducible. Why? Because otherwise you can get a more concise program for its output by computing it first and then running it. Look at this diagram:

$$\text{program}_2 \rightarrow \textbf{Computer} \rightarrow \text{program}_1 \rightarrow \textbf{Computer} \rightarrow \text{output}$$

---

[16]Atoms of thought!

If program$_1$ is as concise as possible, then program$_2$ cannot be much more concise than program$_1$. Why? Well, consider a fixed-sized routine for running a program and then immediately running its output. Then

$$\text{program}_2 + \text{fixed-size routine} \rightarrow \textbf{Computer} \rightarrow \text{output}$$

produces exactly the same output as program$_1$ and would be a more concise program for producing that output than program$_1$ is. But this is impossible because it contradicts our hypothesis that program$_1$ was already as small as possible. *Q.E.D.*

Why should elegant programs interest philosophers? Well, because of Occam's razor, because the best theory to explain a fixed set of data is an elegant program!

But how can we get irreducible truths? Well, just try **proving** that a program is elegant!

VI Irreducible Mathematical Truths. Examples of Logical Irreducibility: Proving a Program is Elegant

Hauptsatz: *You cannot prove that a program is elegant if its size is substantially larger than the size of the algorithm for generating all the theorems in your theory.*

Proof: The basic idea is to run the first provably elegant program you encounter when you systematically generate all the theorems, and that is substantially larger than the size of the algorithm for generating all the theorems. Contradiction, unless no such theorem can be demonstrated, or unless the theorem is false.

Now I'll explain why this works. We are given a formal axiomatic mathematical theory:

$$\text{theory} = \text{program} \rightarrow \textbf{Computer} \rightarrow \text{set of all theorems}$$

We may suppose that this theory is an elegant program, i.e., as concise as possible for producing the set of theorems that it does. Then the size of this program is by definition the complexity of the theory, since it is the size of the smallest program for systematically generating the set of all the theorems, which are all the consequences of the axioms. Now consider a fixed-size routine with the property that

theory + fixed-size routine → **Computer** →
output of the first provably elegant program larger than
complexity of theory

More precisely,

theory + fixed-size routine → **Computer** →
output of the first provably elegant program larger than
(complexity of theory + size of the fixed-size routine)

This proves our assertion that a mathematical theory cannot prove that a program is elegant if that program is substantially larger than the complexity of the theory.

Here is the proof of this result in more detail. The fixed-size routine knows its own size and is given the theory, a computer program for generating theorems, whose size it measures and which it then runs, until the first theorem is encountered asserting that a particular program P is elegant that is larger than the total input to the computer. The fixed-size routine then runs the program P, and finally produces as output the same output as P produces. But this is impossible, because the output from P cannot be obtained from a program that is smaller than P is, not if, as we assume by hypothesis, all the theorems of the theory are true and P is actually elegant. Therefore P cannot exist. In other words, if there is a provably elegant program P whose size is greater than the complexity of the theory + the size of this fixed-size routine, either P is actually inelegant or we have a contradiction. *Q.E.D.*

Because no mathematical theory of finite complexity can enable you to determine all the elegant programs, the following is immediate:

Corollary: *The mathematical universe has infinite complexity.*[17]

This strengthens Gödel's 1931 refutation of Hilbert's belief that a single, fixed formal axiomatic theory could capture all of mathematical truth.

Given the significance of this conclusion, it is natural to demand more information. You'll notice that I never said **which** computer programming language I was using!

Well, you can actually carry out this proof using either high-level languages such as the version of LISP that I use in *The Unknowable,* or using

---

[17]On the other hand, our current mathematical theories are **not** very complex. On pages 773–774 of *A New Kind of Science,* Wolfram makes this point by exhibiting essentially all of the axioms for traditional mathematics—in just two pages! However, a program to generate all the theorems would be larger.

low-level binary machine languages, such as the one that I use in *The Limits of Mathematics.* In the case of a high-level computer programming language, one measures the size of a program in characters (or 8-bit bytes) of text. In the case of a binary machine language, one measures the size of a program in 0/1 bits. My proof works either way.

But I must confess that not all programming languages permit my proof to work out this neatly. The ones that do are the kinds of programming languages that you use in AIT, the ones for which program-size complexity has elegant properties instead of messy ones, the ones that directly expose the fundamental nature of this complexity concept (which is also called algorithmic information content), not the programming languages that bury the basic idea in a mass of messy technical details.

This paper started with philosophy, and then we developed a mathematical theory. Now let's go back to philosophy. In the last three sections of this paper we'll discuss the philosophical implications of AIT.

VII Could We Ever Be Sure that We Had the Ultimate TOE? [Barrow 1995]

"The search for a 'Theory of Everything' is the quest for an ultimate compression of the world. Interestingly, Chaitin's proof of Gödel's incompleteness theorem using the concepts of complexity and compression reveals that Gödel's theorem is equivalent to the fact that one cannot prove a sequence to be incompressible. We can never prove a compression to be the ultimate one; there might be a yet deeper and simpler unification waiting to be found."

—John Barrow, essay on "Theories of Everything" in Cornwell, *Nature's Imagination,* 1995, reprinted in Barrow, *Between Inner Space and Outer Space,* 1999.

Here is the first philosophical application of AIT. According to astrophysicist John Barrow, my work implies that even if we had the optimum, perfect, minimal (elegant!) TOE, we could never be sure a simpler theory would not have the same explanatory power.

("Explanatory power" is a pregnant phrase, and one can make a case that it is a better name to use than the dangerous word "complexity", which has many other possible meanings. One could then speak of a theory with N bits of algorithmic explanatory power, rather than describe it as a theory having a program-size complexity of N bits. [Françoise Chaitin-Chatelin, private communication])

Well, you can dismiss Barrow by saying that the idea of having the ultimate TOE is pretty crazy—who expects to be able to read the mind of God?! Actually, Wolfram believes that a systematic computer search might well find the ultimate TOE.[18] I hope he continues working on this project!

In fact, Wolfram thinks that he not only might be able to find the ultimate TOE, he might even be able to show that it is the simplest possible TOE! How does he escape the impact of my results? Why doesn't Barrow's observation apply here?

First of all, Wolfram is not very interested in proofs, he prefers computational evidence. Second, Wolfram does not use program-size complexity as his complexity measure. He uses much more down-to-earth complexity measures. Third, he is concerned with extremely simple systems, while my methods apply best to objects with high complexity.

Perhaps the best way to explain the difference is to say that he is looking at "hardware" complexity, and I'm looking at "software" complexity. The objects he studies have complexity less than or equal to that of a universal computer. Those I study have complexity much larger than a universal computer. For Wolfram, a universal computer is the maximum possible complexity, and for me it is the minimum possible complexity.

Anyway, now let's see what's the message from AIT for the working mathematician.

VIII Should Mathematics Be More Like Physics? Must Mathematical Axioms Be Self-Evident?

"A deep but easily understandable problem about prime numbers is used in the following to illustrate the parallelism between the heuristic reasoning of the mathematician and the inductive reasoning of the physicist... [M]athematicians and physicists think alike; they are led, and sometimes misled, by the same patterns of plausible reasoning."

—George Pólya, "Heuristic Reasoning in the Theory of Numbers", 1959, reprinted in Alexanderson, *The Random Walks of George Pólya*, 2000.

"The role of heuristic arguments has not been acknowledged in the philosophy of mathematics, despite the crucial role that they play in mathematical discovery. The mathematical notion of proof is strikingly at variance with the notion of proof in other

---

[18]See pages 465–471, 1024–1027 of *A New Kind of Science*.

areas... Proofs given by physicists do admit degrees: of two proofs given of the same assertion of physics, one may be judged to be more correct than the other."

—Gian-Carlo Rota, "The Phenomenology of Mathematical Proof", 1997, reprinted in Jacquette, *Philosophy of Mathematics,* 2002, and in Rota, *Indiscrete Thoughts,* 1997.

"There are two kinds of ways of looking at mathematics... the Babylonian tradition and the Greek tradition... Euclid discovered that there was a way in which all the theorems of geometry could be ordered from a set of axioms that were particularly simple... The Babylonian attitude... is that you know all of the various theorems and many of the connections in between, but you have never fully realized that it could all come up from a bunch of axioms... [E]ven in mathematics you can start in different places... In physics we need the Babylonian method, and not the Euclidian or Greek method."

—Richard Feynman, *The Character of Physical Law,* 1965, Chapter 2, "The Relation of Mathematics to Physics".

"The physicist rightly dreads precise argument, since an argument which is only convincing if precise loses all its force if the assumptions upon which it is based are slightly changed, while an argument which is convincing though imprecise may well be stable under small perturbations of its underlying axioms."

—Jacob Schwartz, "The Pernicious Influence of Mathematics on Science", 1960, reprinted in Kac, Rota, Schwartz, *Discrete Thoughts,* 1992.

"It is impossible to discuss realism in logic without drawing in the empirical sciences... A truly realistic mathematics should be conceived, in line with physics, as a branch of the theoretical construction of the one real world and should adopt the same sober and cautious attitude toward hypothetic extensions of its foundation as is exhibited by physics."

—Hermann Weyl, *Philosophy of Mathematics and Natural Science,* 1949, Appendix A, "Structure of Mathematics", p. 235.

The above quotations are eloquent testimonials to the fact that although mathematics and physics are different, maybe they are not **that** different! Admittedly, math organizes our mathematical experience, which is mental or computational, and physics organizes our physical experience.[19] They are certainly not exactly the same, but maybe it's a matter of degree, a continuum of possibilities, and not an absolute, black and white difference.

Certainly, as both fields are currently practiced, there is a definite difference in **style**. But that could change, and is to a certain extent a matter of fashion, not a fundamental difference.

A good source of essays that I—but perhaps not the authors!—regard as generally supportive of the position that math be considered a branch of physics is Tymoczko, *New Directions in the Philosophy of Mathematics,* 1998. In particular there you will find an essay by Lakatos giving the name "quasi-empirical" to this view of the nature of the mathematical enterprise.

---

[19] And in physics everything is an approximation, no equation is exact.

Why is my position on math "quasi-empirical"? Because, as far as I can see, this is the only way to accommodate the existence of irreducible mathematical facts gracefully. Physical postulates are never self-evident, they are justified pragmatically, and so are close relatives of the not at all self-evident irreducible mathematical facts that I exhibited in Section VI.

I'm not proposing that math is a branch of physics just to be controversial. I was forced to do this against my will! This happened in spite of the fact that I'm a mathematician and I love mathematics, and in spite of the fact that I started with the traditional Platonist position shared by most working mathematicians. I'm proposing this because I want mathematics to work better and be more productive. Proofs are fine, but if you can't find a proof, you should go ahead using heuristic arguments and conjectures.

Wolfram's *A New Kind of Science* also supports an experimental, quasi-empirical way of doing mathematics. This is partly because Wolfram is a physicist, partly because he believes that unprovable truths are the rule, not the exception, and partly because he believes that our current mathematical theories are highly arbitrary and contingent. Indeed, his book may be regarded as a very large chapter in experimental math. In fact, he had to develop his own programming language, *Mathematica,* to be able to do the massive computations that led him to his conjectures.

See also Tasić, *Mathematics and the Roots of Postmodern Thought,* 2001, for an interesting perspective on intuition versus formalism. This is a key question—indeed in my opinion it's an inescapable issue—in any discussion of how the game of mathematics should be played. And it's a question with which I, as a working mathematician, am passionately concerned, because, as we discussed in Section VI, formalism has severe limitations. Only intuition can enable us to go forward and create new ideas and more powerful formalisms.

And what are the wellsprings of mathematical intuition and creativity? In his important forthcoming book on creativity, Tor Nørretranders makes the case that a peacock, an elegant, graceful woman, and a beautiful mathematical theory, are all shaped by the same forces, namely what Darwin referred to as "sexual selection". Hopefully this book will be available soon in a language other than Danish! Meanwhile, see my dialogue with him in my book *Conversations with a Mathematician.*[20]

---

[20][An English edition of Nørretranders' book on creativity is now available: *The Generous Man,* Thunder's Mouth Press, 2005.]

Now, for our last topic, let's look at the entire physical universe!

IX Is the Universe Like π or Like Ω? Reason versus Randomness! [Brisson, Meyerstein 1995]

"Parce qu'on manquait d'une définition rigoreuse de complexité, celle qu'a proposée la TAI [théorie algorithmique de l'information], confondre π avec Ω a été plutôt la règle que l'exception. Croire, parce que nous avons ici affaire à une croyance, que toutes les suites, puisqu'elles ne sont que l'enchaînement selon une règle rigoureuse de symboles déterminés, peuvent toujours être comprimées en quelque chose de plus simple, voilà la source de l'erreur du réductionnisme. Admettre la complexité a toujours paru insupportable aux philosophes, car c'était renoncer à trouver un sens rationnel à la vie des hommes."

—Brisson, Meyerstein, *Puissance et Limites de la Raison,* 1995, "Postface. L'erreur du réductionnisme", p. 229.

First let me explain what the number Ω is. It's the jewel in AIT's crown, and it's a number that has attracted a great deal of attention, because it's a very **dangerous** number! Ω is defined to be the halting probability of what computer scientists call a universal computer, or universal Turing machine.[21] So Ω is a probability and therefore it's a real number, a number measured with infinite precision, that's between zero and one.[22] That may not sound too dangerous!

What's dangerous about Ω is that (a) it has a simple, straightforward mathematical definition, but at the same time (b) its numerical value is maximally unknowable, because a formal mathematical theory whose program-size complexity or explanatory power is N bits cannot enable you to determine more than N bits of the base-two expansion of Ω! In other words, if you want to calculate Ω, theories don't help very much, since it takes N bits of theory to get N bits of Ω. In fact, the base-two bits of Ω are maximally complex, there's no redundancy, and Ω is the prime example of how unadulterated infinite complexity arises in pure mathematics!

How about $\pi = 3.1415926\ldots$ the ratio of the circumference of a circle to its diameter? Well, π **looks** pretty complicated, pretty lawless. For example,

---

[21]In fact, the precise value of Ω actually depends on the choice of computer, and in *The Limits of Mathematics* I've done that, I've picked one out.

[22]It's ironic that the star of a discrete theory is a real number! This illustrates the creative tension between the continuous and the discrete.

all its digits seem to be equally likely,[23] although this has never been proven.[24] If you are given a bunch of digits from deep inside the decimal expansion of π, and you aren't told where they come from, there doesn't seem to be any redundancy, any pattern. But of course, according to AIT, π in fact only has **finite** complexity, because there are algorithms for calculating it with arbitrary precision.[25]

Following Brisson, Meyerstein, *Puissance et Limites de la Raison*, 1995, let's now finally discuss whether the physical universe is like $\pi = 3.1415926\ldots$ which only has a finite complexity, namely the size of the smallest program to generate π, or like Ω, which has unadulterated infinite complexity. Which is it?!

Well, if you believe in quantum physics, then Nature plays dice, and that generates complexity, an infinite amount of it, for example, as frozen accidents, mutations that are preserved in our DNA. So at this time most scientists would bet that the universe has infinite complexity, like Ω does. But then the world is incomprehensible, or at least a large part of it will always remain so, the accidental part, all those frozen accidents, the contingent part.

But some people still hope that the world has finite complexity like π, it just **looks** like it has high complexity. If so, then we might eventually be able to comprehend everything, and there is an ultimate TOE! But then you have to believe that quantum mechanics is wrong, as currently practiced, and that all that quantum randomness is really only **pseudo-randomness**, like what you find in the digits of π. You have to believe that the world is actually deterministic, even though our current scientific theories say that it isn't!

I think Vienna physicist Karl Svozil feels that way [private communication; see his *Randomness & Undecidability in Physics*, 1994]. I know Stephen Wolfram does, he says so in his book. Just take a look at the discussion of fluid turbulence and of the second law of thermodynamics in *A New Kind of Science*. Wolfram believes that very simple deterministic algorithms ultimately account for all the apparent complexity we see around us, just like

---

[23]In any base all the digits of Ω are equally likely. This is called "Borel normality". For a proof, see my book *Exploring Randomness*. For the latest on Ω, see Calude, *Information and Randomness*.

[24]Amazingly enough, there's been some recent progress in this direction by Bailey and Crandall.

[25]In fact, some terrific new ways to calculate π have been discovered by Bailey, Borwein and Plouffe. π lives, it's not a dead subject!

they do in π.[26] He believes that the world **looks** very complicated, but is actually very simple. There's no randomness, there's only pseudo-randomness. Then nothing is contingent, everything is necessary, everything happens for a reason. [Leibniz!]

Who knows! Time will tell!

Or perhaps from **inside** this world we will never be able to tell the difference, only an **outside** observer could do that [Svozil, private communication].

Postscript

Readers of this paper may enjoy the somewhat different perspective in my chapter "Complexité, logique et hasard" in Benkirane, *La Complexité*. Leibniz is there too.

In addition, see my *Conversations with a Mathematician*, a book on philosophy disguised as a series of dialogues—not the first time that this has happened!

Last but not least, see Zwirn, *Les Limites de la Connaissance*, that also supports the thesis that understanding is compression, and the masterful multi-author two-volume work, *Kurt Gödel, Wahrheit & Beweisbarkeit*, a treasure trove of information about Gödel's life and work.

Acknowledgement

Thanks to Tor Nørretranders for providing the original German for the Einstein quotation at the beginning of this paper, and also the word for word translation.

The author is grateful to Françoise Chaitin-Chatelin for innumerable stimulating philosophical discussions. He dedicates this paper to her unending quest to understand.

Bibliography

• Gerald W. Alexanderson, *The Random Walks of George Pólya*, MAA, 2000.

---

[26]In fact, Wolfram himself explicitly makes the connection with π. See **meaning of the universe** on page 1027 of *A New Kind of Science*.

- John D. Barrow, Frank J. Tipler, *The Anthropic Cosmological Principle*, Oxford University Press, 1986.

- John D. Barrow, *Between Inner Space and Outer Space*, Oxford University Press, 1999.

- Reda Benkirane, *La Complexité, Vertiges et Promesses*, Le Pommier, 2002.

- Max Born, *Experiment and Theory in Physics*, Cambridge University Press, 1943. Reprinted by Dover, 1956.

- Luc Brisson, F. Walter Meyerstein, *Inventer l'Univers*, Les Belles Lettres, 1991.

- Luc Brisson, F. Walter Meyerstein, *Inventing the Universe*, SUNY Press, 1995.

- Luc Brisson, F. Walter Meyerstein, *Puissance et Limites de la Raison*, Les Belles Lettres, 1995.

- F. Bródy, T. Vámos, *The Neumann Compendium*, World Scientific, 1995.

- Bernd Buldt et al., *Kurt Gödel, Wahrheit & Beweisbarkeit. Band 2: Kompendium zum Werk,* öbv & hpt, 2002.

- Cristian S. Calude, *Information and Randomness*, Springer-Verlag, 2002.

- Gregory J. Chaitin, *The Limits of Mathematics, The Unknowable, Exploring Randomness, Conversations with a Mathematician*, Springer-Verlag, 1998, 1999, 2001, 2002.

- John Cornwell, *Nature's Imagination*, Oxford University Press, 1995.

- COSRIMS, *The Mathematical Sciences*, MIT Press, 1969.

- Albert Einstein, *Ideas and Opinions*, Crown, 1954. Reprinted by Modern Library, 1994.

- Albert Einstein, *Autobiographical Notes*, Open Court, 1979.

- Richard Feynman, *The Character of Physical Law*, MIT Press, 1965. Reprinted by Modern Library, 1994, with a thoughtful introduction by James Gleick.

- Richard P. Feynman, Robert B. Leighton, Matthew Sands, *The Feynman Lectures on Physics*, Addison-Wesley, 1963.

- Dale Jacquette, *Philosophy of Mathematics*, Blackwell, 2002.

- Mark Kac, Gian-Carlo Rota, Jacob T. Schwartz, *Discrete Thoughts*, Birkhäuser, 1992.

- Eckehart Köhler et al., *Kurt Gödel, Wahrheit & Beweisbarkeit. Band 1: Dokumente und historische Analysen,* öbv & hpt, 2002.

- Bernd-Olaf Küppers, *Information and the Origin of Life*, MIT Press, 1990.

- G. W. Leibniz, *Philosophical Essays*, edited and translated by Roger Ariew and Daniel Garber, Hackett, 1989.

- Ernst Mach, *The Science of Mechanics,* Open Court, 1893.

- Paolo Mancosu, *From Brouwer to Hilbert,* Oxford University Press, 1998.

- Karl Menger, *Reminiscences of the Vienna Circle and the Mathematical Colloquium,* Kluwer, 1994.

- James R. Newman, *The World of Mathematics,* Simon and Schuster, 1956. Reprinted by Dover, 2000.

- Karl R. Popper, *The Logic of Scientific Discovery,* Hutchinson Education, 1959. Reprinted by Routledge, 1992.

- Gian-Carlo Rota, *Indiscrete Thoughts,* Birkhäuser, 1997.

- Paul Arthur Schilpp, *Albert Einstein, Philosopher-Scientist,* Open Court, 1949.

- Karl Svozil, *Randomness & Undecidability in Physics,* World Scientific, 1994.

- Vladimir Tasić, *Mathematics and the Roots of Postmodern Thought,* Oxford University Press, 2001.

- Thomas Tymoczko, *New Directions in the Philosophy of Mathematics*, Princeton University Press, 1998.

- Hermann Weyl, *The Open World,* Yale University Press, 1932. Reprinted by Ox Bow Press, 1989.

- Hermann Weyl, *Philosophy of Mathematics and Natural Science,* Princeton University Press, 1949.

- Stephen Wolfram, *A New Kind of Science,* Wolfram Media, 2002.

- Hervé Zwirn, *Les Limites de la Connaissance,* Odile Jacob, 2000.

Leibniz, information, math & physics

The information-theoretic point of view proposed by Leibniz in 1686 and developed by algorithmic information theory (AIT) suggests that mathematics and physics are not that different. This will be a first-person account of some doubts and speculations about the nature of mathematics that I have entertained for the past three decades, and which have now been incorporated in a digital philosophy paradigm shift that is sweeping across the sciences.

1. What is algorithmic information theory?

The starting point for my own work on AIT forty years ago was the insight that a scientific theory is a computer program that calculates the observations, and that the smaller the program is, the better the theory. If there is no theory, that is to say, no program substantially smaller than the data itself, considering them both to be finite binary strings, then the observations are algorithmically random, theory-less, unstructured, incomprehensible and irreducible.

theory = program $\longrightarrow$ $\boxed{\textbf{Computer}}$ $\longrightarrow$ output = experimental data

So this led me to a theory of randomness based on program-size complexity [1], whose main application turned out to be not in science, but in mathematics, more specifically, in meta-mathematics, where it yields powerful new information-theoretic versions of Gödel's incompleteness theorem [2, 3, 4]. (I'll discuss this in Section 3.)

And from this new information-theoretic point of view, math and physics do not seem too different. In both cases understanding is compression, and is measured by the extent to which empirical data and mathematical theorems are respectively compressed into concise physical laws or mathematical axioms, both of which are embodied in computer software [5].

And why should one use reasoning at all in mathematics?! Why not proceed entirely empirically, more or less as physicists do? Well, the advantange of proving things is that assuming a few bits of axioms is less risky than assuming many empirically-suggested mathematical assertions. (The disadvantage, of course, is the length of the proofs and the risk of faulty proofs.) Each bit in an irreducible axiom of a mathematical theory is a freely-chosen independent assumption, with an *a priori* probability of half of being the right choice, so one wants to reduce the number of such independent choices to a minimum in creating a new theory.

So this point of view would seem to suggest that while math and physics are admittedly different, perhaps they are not as different as most people usually believe. Perhaps we should feel free to pursue not only rigorous, formal modern proofs, but also the swash-buckling experimental math that Euler enjoyed so much. And in fact theoretical computer scientists have to some extent already done this, since their $\mathbf{P} \neq \mathbf{NP}$ hypothesis is probably currently the best candidate for canonization as a new axiom. And, as is suggested in [6], another possible candidate is the Riemann hypothesis.

But before discussing this in more detail, I'd like to tell how I discovered that in 1686 Leibniz anticipated some of the basic ideas of AIT.

2. How Leibniz almost invented algorithmic information theory [7]

One day last year, while preparing my first philosophy paper [5], for a philosophy congress in Bonn, I was reading a little book on philosophy by Hermann Weyl that was published in 1932, and I was amazed to find the following, which captures the essential idea of my definition of algorithmic randomness:

> "The assertion that nature is governed by strict laws is devoid of all content if we do not add the statement that it is governed by mathematically simple laws... That **the notion of law**

becomes empty when an arbitrary complication is per-mitted was already pointed out by Leibniz in his *Metaphysical Treatise* [*Discourse on Metaphysics*]. Thus simplicity becomes a working principle in the natural sciences."

—Weyl [8, pp. 40–42]. See a similar discussion on pp. 190–191 of Weyl [9], Section 23A, "Causality and Law".

In fact, I actually read Weyl [9] as a teenager, before inventing AIT at age 15, but the matter is not stated so sharply there. And a few years ago I stumbled on the above-quoted text in Weyl [8], but hadn't had the time to pursue it until stimulated to do so by an invitation from the German Philosophy Association to talk at their 2002 annual congress, that happened to be on limits and how to transcend them.

So I got a hold of Leibniz's *Discourse on Metaphysics* to see what he actually said. Here it is:

"As for the simplicity of the ways of God, this holds properly with respect to his means, as opposed to the variety, richness, and abundance, which holds with respect to his ends or effects."

"...not only does nothing completely irregular occur in the world, but we would not even be able to imagine such a thing. Thus, let us assume, for example, that someone jots down a number of points at random on a piece of paper, as do those who practice the ridiculous art of geomancy.[1] I maintain that it is possible to find a geometric line whose [m]otion is constant and uniform, following a certain rule, such that this line passes through all the points in the same order in which the hand jotted them down."

"But, **when a rule is extremely complex, what is in conformity with it passes for irregular**. Thus, one can say, in whatever manner God might have created the world, it would always have been regular and in accordance with a certain general order. But **God has chosen the most perfect world, that is, the one which is at the same time the simplest in hypotheses and the richest in phenomena**, as might be a line in geometry whose construction is easy and whose properties and effects are extremely remarkable and widespread."

---

[1][A way to foretell the future; a form of divination.]

—Leibniz, *Discourse on Metaphysics,* 1686, Sections 5–6, as translated by Ariew and Garber [10, pp. 38–39].

$$\text{ideas} = \text{input} \longrightarrow \boxed{\textbf{Mind of God}} \longrightarrow \text{output} = \text{the universe}$$

And after finishing my paper [5] for the Bonn philosophy congress, I learned that Leibniz's original *Discourse on Metaphysics* was in French, which I know, and fortunately not in Latin, which I don't know, and that it was readily available from France:

"Pour ce qui est de la simplicité des voyes de Dieu, elle a lieu proprement à l'égard des moyens, comme au contraire la varieté, richesse ou abondance y a lieu à l'égard des fins ou effects."

"...non seulement rien n'arrive dans le monde, qui soit absolument irregulier, mais on ne sçauroit mêmes rien feindre de tel. Car supposons par exemple que quelcun fasse quantité de points sur le papier à tout hazard, comme font ceux qui exercent l'art ridicule de la Geomance, je dis qu'il est possible de trouver une ligne geometrique dont la [m]otion soit constante et uniforme suivant une certaine regle, en sorte que cette ligne passe par tous ces points, et dans le même ordre que la main les avoit marqués."

"Mais **quand une regle est fort composée, ce qui luy est conforme, passe pour irrégulier**. Ainsi on peut dire que de quelque maniere que Dieu auroit créé le monde, il auroit tousjours esté regulier et dans un certain ordre general. Mais **Dieu a choisi celuy qui est** le plus parfait, c'est à dire celuy qui est en même temps **le plus simple en hypotheses et le plus riche en phenomenes,** comme pourroit estre une ligne de Geometrie dont la construction seroit aisée et les proprietés et effects seroient fort admirables et d'une grande étendue."

—Leibniz, *Discours de métaphysique,* **V–VI** [11, pp. 40–41].

(Here "dont la motion" is my correction. The Gallimard text [11] states "dont la notion," an obvious misprint, which I've also corrected in the English translation by Ariew and Garber.)

So, in summary, Leibniz observes that for any finite set of points there is a mathematical formula that produces a curve that goes through them all, and it can be parametrized so that it passes through the points in the order that they were given and with a constant speed. So this cannot give us a definition of what it means for a set of points to obey a law. But if the formula is very simple, and the data is very complex, then that's a *real law*!

Recall that Leibniz was at the beginning of the modern era, in which ancient metaphysics was colliding with modern empirical science. And he was a great mathematician as well as a philosopher. So here he is able to take a stab at clarifying what it means to say that Nature is lawful and what are the conditions for empirical science to be possible.

AIT puts more meat on Leibniz's proposal, it makes his ideas more precise by giving a precise definition of complexity.

And AIT goes beyond Leibniz by using program-size complexity to clarify what it means for a sequence of observations to be lawless, one which has no theory, and by applying this to studying the limits of formal axiomatic reasoning, i.e., what can be achieved by mindlessly and mechanically grinding away deducing all possible consequences of a fixed set of axioms. (I'll say more about metamathematical applications of AIT in Section 3 below.)

$$\text{axioms = program} \longrightarrow \boxed{\textbf{Computer}} \longrightarrow \text{output = theorems}$$

By the way, the articles by philosophy professors that I've seen that discuss the above text by Leibniz criticize what they see as the confused and ambiguous nature of his remarks. On the contrary, I admire his prescience and the manner in which he has unerringly identified the central issue, the key idea. He even built a mechanical calculator and with his speculations regarding a *Characteristica Universalis* ("Adamic" language of creation) envisioned something that Martin Davis [12] has argued was a direct intellectual ancestor of the universal Turing machine, which is precisely the device that is needed in order for AIT to be able to quantify Leibniz's original insight!

Davis quotes some interesting remarks by Leibniz about the practical utility of his calculating machine. Here is part of the Davis Leibniz quote:

> "And now that we may give final praise to the machine we may say that it will be desirable to all who are engaged in computations which, it is well known, are the managers of financial affairs, the administrators of others' estates, merchants, surveyors, geographers, navigators, astronomers... For it is unworthy of excellent men to lose hours like slaves in the labor of calculations

> which could safely be relegated to anyone else if the machine were used."

This reminds me of a transcript of a lecture that von Neumann gave at the inauguration of the NORC (Naval Ordnance Computer) that I read many years ago. It attempted to convince people that computers were of value. It was a hard sell! The obvious practical and scientific utility of calculators and computers, though it was evident to Leibniz, Babbage and von Neumann, was far from evident to most people. Even von Neumann's colleagues at the Princeton Institute for Advanced Study completely failed to understand this (see Casti [13]).

And I am almost forgetting something important that I read in E. T. Bell [14] as a child, which is that **Leibniz invented base-two binary notation** for integers. Bell reports that this was a result of Leibniz's interest in Chinese culture; no doubt he got it from the *I Ching*. So in a sense, all of information theory derives from Leibniz, for he was the first to emphasize the creative combinatorial potential of the 0 and 1 bit, and how everything can be built up from this one elemental choice, from these two elemental possibilities. So, perhaps not entirely seriously, I should propose changing the name of the unit of information from the *bit* to the *leibniz*!

3. The halting probability Ω and information-theoretic incompleteness

Enough philosophy, let's do some mathematics! The first step is to pick a universal binary computer U with the property that for any other binary computer C there is a binary prefix π_C such that

$$U(\pi_C\, p) = C(p).$$

Here p is a binary program for C and the prefix π_C tells U how to *simulate* C and does not depend on p. In the U that I've picked, π_C consists of a description of C written in the high-level non-numerical functional programming language LISP, which is much like a computerized version of set theory, except that all sets are finite.

Next we define the *algorithmic information content* (program-size complexity) of a LISP symbolic expression (S-expression) X to be the size in bits

$|p|$ of the smallest binary program p that makes our chosen U compute X:

$$H(X) \equiv \min_{U(p)=X} |p|.$$

Similarly, the information content or complexity of a formal axiomatic theory with the infinite set of theorems T is defined to be the size in bits of the smallest program that makes U generate the infinite set of theorems T, which is a set of S-expressions.

$$H(T) \equiv \min_{U(p)=T} |p|.$$

Think of this as the minimum number of bits required to tell U how to run through all possible proofs and systematically generate all the consequences of the fixed set of axioms. $H(T)$ is the size in bits of the most concise axioms for T.

Next we define the celebrated halting probability Ω:

$$\Omega \equiv \sum_{U(p) \text{ halts}} 2^{-|p|}.$$

A small technical detail: To get this sum to converge it is necessary that programs for U be "self-delimiting." I.e., no extension of a valid program is a valid program, the set of valid programs has to be a prefix-free set of bit strings.

So Ω is now a specific, well-defined real number between zero and one, and let's consider its binary expansion, i.e., its base-two representation. Discarding the initial decimal (or binary) point, that's an infinite binary sequence $b_1 b_2 b_3 \ldots$ To eliminate any ambiguity in case Ω should happen to be a dyadic rational (which it actually isn't), let's agree to change $1000\ldots$ to $0111\ldots$ here if necessary.

Right away we get into trouble. From the fact that knowing the first N bits of Ω

$$\Omega_N \equiv b_1 b_2 b_3 \ldots b_N$$

would enable us to answer the halting problem for every program p for U with $|p| \leq N$, it is easy to see that the bits of Ω are *computationally irreducible*:

$$H(\Omega_N) \geq N - c.$$

And from this it follows using a straight-forward program-size argument (see [3]) that the bits of Ω are also *logically irreducible*.

What does this mean? Well, consider a formal axiomatic theory with theorems T, an infinite set of S-expressions. If we assume that a theorem of the form "The kth bit of Ω is 0/1" is in T only if it's true, then T cannot enable us to determine more than $H(T) + c'$ bits of Ω.

So the bits of Ω are irreducible mathematical facts, they are mathematical facts that contradict Leibniz's principle of sufficient reason by **being true for no reason**. They must, to use Kantian terminology, be apprehended as things in themselves. They cannot be deduced as consequences of any axioms or principles that are simpler than they are.

(By the way, this also implies that the bits of Ω are statistically random, e.g., Ω is absolutely Borel normal in every base. I.e., all blocks of digits of the same size have equal limiting relative frequency, regardless of the radix chosen for representing Ω.)

Furthermore, in my 1987 Cambridge University Press monograph [15] I celebrate the fact that the bits of Ω can be encoded via a diophantine equation. There I exhibit an exponential diophantine equation $L(k, \mathbf{x}) = R(k, \mathbf{x})$ with parameter k and about twenty-thousand unknowns $\mathbf{x}$ that has infinitely many solutions iff the kth bit of Ω is a 1. And recently Ord and Kieu [16] have shown that this can also be accomplished using the even/odd parity of the number of solutions, rather than its finite/infinite cardinality. So Ω's irreducibility also infects elementary number theory!

These rather brutal incompleteness results show how badly mistaken Hilbert was to assume that a fixed formal axiomatic theory could encompass all of mathematics. And if you have to extend the foundations of mathematics by constantly adding new axioms, new concepts and fundamental principles, then mathematics becomes much more tentative and begins to look much more like an empirical science. At least I think so, and you can even find quotes by Gödel that I think point in the same direction.

These ideas are of course controversial; see for example a highly critical review of two of my books in the *AMS Notices* [17]. I discuss the hostile reaction of the logic community to my ideas in more detail in an interview with performance artist Marina Abramovic [18]. Here, however, I prefer to tell why I think that the world is actually moving rather quickly in my direction. In fact, I believe that my ideas are now part of an unstoppable tidal wave of change spreading across the sciences!

4. The digital philosophy paradigm shift

As I have argued in the second half of my 2002 paper in the *EATCS Bulletin* [19], what we are witnessing now is a dramatic convergence of mathematics with theoretical computer science and with theoretical physics. The participants in this paradigm shift believe that *information* and *computation* are fundamental concepts in all three of these domains, and that what physical systems actually do is computation, i.e., information processing. In other words, as is asked on the cover of a recent issue of *La Recherche* with an article [20] about this, "Is God a Computer?"

But that is not quite right. Rather, we should ask, "Is God a Programmer?" The intellectual legacy of the West, and in this connection let me recall Pythagoras, Plato, Galileo and James Jeans, states that "Everything is number; God is a mathematician." We are now beginning to believe something slightly different, a refinement of the original Pythagorean credo: "Everything is software; God is a computer programmer." Or perhaps I should say: "All is algorithm!" Just as DNA programs living beings, God programs the universe.

In the digital philosophy movement I would definitely include: the extremely active field of quantum information and quantum computation [21], Wolfram's work [22] on *A New Kind of Science*, Fredkin's work on reversible cellular automata and his website at `http://digitalphilosophy.org` (the pregnant phrase "digital philosophy" is due to Fredkin), the Bekenstein-t'Hooft "holographic principle" [23], and AIT. Ideas from theoretical physics and theoretical computer science are definitely leaking across the traditional boundaries between these two fields. And this holds for AIT too, because its two central concepts are versions of *randomness* and of *entropy*, which are ideas that I took with me from physics and into mathematical logic.

Wolfram's work is particularly relevant to our discussion of the nature of mathematics, because he believes that most simple systems are either trivial or equivalent to a universal computer, and therefore that mathematical questions are either trivial or can never be solved, except, so to speak, for a set of measure zero. This he calls his *principle of computational equivalence*, and it leads him to take the incompleteness phenomenon much more seriously than most mathematicians do. In line with his thesis, his book presents a great deal of computational evidence, but not many proofs.

Another important issue studied in Wolfram's book [22] is the question of whether, to use Leibnizian terminology, mathematics is necessary or is contingent. I.e., would intelligent creatures on another planet necessarily discover the same concepts that we have, or might they develop a perfectly viable mathematics that we would have a great deal of trouble in recognizing as such? Wolfram gives a number of examples that suggest that the latter is in fact the case.

I should also mention some recent books on the quasi-empirical view of mathematics [24] and on experimental mathematics [25, 26], as well as Douglas Robertson's two volumes [27, 28] on information as a key historical and cultural parameter and motor of social change, and John Maynard Smith's related books on biology [29, 30].

Maynard Smith and Szathmáry [29, 30] measure biological evolutionary progress in terms of abrupt improvements in the way information is represented and transmitted inside living organisms. Robertson sees social evolution as driven by the same motor. According to Robertson [27, 28], spoken language defines the human, writing creates civilization, the printing press provoked the Renaissance, and the Internet is weaving a new World-Wide Web. These are abrupt improvements in the way human society is able to store and transmit information. And they result in abrupt increases in cultural complexity, in abrupt increases in social *intelligence,* as it were.

(And for the latest results on Ω, see Calude [31].)

5. Digital philosophy is Leibnizian; Leibniz's legacy

None of us who made this paradigm shift happen were students of Leibniz, but he anticipated us all. As I hinted in a letter to *La Recherche,* in a sense all of Wolfram's thousand-page book is the development of *one sentence* in Leibniz:

> **"Dieu a choisi celuy qui est... le plus simple en hypotheses et le plus riche en phenomenes"**
> [God has chosen that which is the most simple in hypotheses and the most rich in phenomena]

This presages Wolfram's basic insight that simple programs can have very complicated-looking output.

And all of *my* work may be regarded as the development of another sentence in Leibniz:

> **"Mais quand une regle est fort composée, ce qui luy est conforme, passe pour irrégulier"**
> [But when a rule is extremely complex, that which conforms to it passes for random]

Here I see the germ of my definition of algorithmic randomness and irreducibility.

Newtonian physics is now receding into the dark, distant intellectual past. It's not just that it has been superseded by quantum physics. No, it's much deeper than that. In our new interest in **complex** systems, the concepts of energy and matter take second place to the concepts of information and computation. And the continuum mathematics of Newtonian physics now takes second place to the combinatorial mathematics of complex systems.

As E. T. Bell stated so forcefully [32], Newton made one big contribution to math, involving the continuum, but Leibniz made *two*: his work on the continuum and his work on discrete combinatorics (which Leibniz named). Newton obliterated Leibniz and stole from him both his royal patron and the credit for the calculus. Newton was buried with full honors at Westminster Abbey, while a forgotten Leibniz was accompanied to his grave by only his secretary. But, as E. T. Bell stated a half a century ago [32], with every passing year, the shadow cast by Leibniz gets larger and larger.

How right Bell was! The digital philosophy paradigm is a direct intellectual descendent of Leibniz, it is part of the Leibnizian legacy. The human race has finally caught up with this part of Leibniz's thinking. Are there, Wolfram and I wonder, more treasures there that we have not yet been able to decipher and appreciate?

6. Acknowledgment; Coda on the continuum and the Kabbalah

The author wishes to thank Françoise Chaitin-Chatelin for sharing with him her understanding and appreciation of Leibniz, during innumerable lengthy conversations. In her opinion, however, this essay does Leibniz an injustice by completely ignoring his deep interest in the "labyrinth of the continuum," which is her specialty.

Let me address her concern. According to Leibniz, the integers are human, the discrete is at the level of Man. But the continuum transcends

Man and brings us closer to God. Indeed, Ω is transcendent, and may be regarded as the concentrated essence of mathematical creativity. In a note on the Kabbalah, which regards Man as perfectable and evolving towards God, Leibniz [33, pp. 112–115] observes that with time we shall know all interesting theorems with proofs of up to any given fixed size, and this can be used to measure human progress.

If the axioms and rules of inference are fixed, then this kind of progress can be achieved mechanically by brute force, which is not very interesting. The interesting case is allowing **new** axioms and concepts. So I would propose instead that human progress—purely intellectual, not **moral** progress—be measured by the number of bits of Ω that we have been able to determine up to any given time.

Let me end with Leibniz's remarks about the effects of this kind of progress [33, pp. 115]:

> "If this happens, it must follow that those minds which are not yet sufficiently capable will become more capable so that they can comprehend and invent such great theorems, which are necessary to understand nature more deeply and to reduce physical truths to mathematics, for example, to understand the mechanical functioning of animals, to forsee certain future contingencies with a certain degree of accuracy, and to do certain wonderful things in nature, which are now beyond our capacity…"

> "Every mind has a horizon in respect to its present intellectual capacity but not in respect to its future intellectual capacity."

References

[1] G. J. Chaitin, *Exploring Randomness,* Springer-Verlag, 2001.

[2] G. J. Chaitin, *The Unknowable,* Springer-Verlag, 1999.

[3] G. J. Chaitin, *The Limits of Mathematics,* Springer-Verlag, 1998.

[4] G. J. Chaitin, *Conversations with a Mathematician,* Springer-Verlag, 2002.

[5] G. J. Chaitin, "On the intelligibility of the universe and the notions of simplicity, complexity and irreducibility," German Philosophy Association, in press.

[6] M. du Sautoy, *The Music of the Primes,* HarperCollins, 2003.

[7] G. J. Chaitin, *From Philosophy to Program Size,* Tallinn Cybernetics Institute, in press.

[8] H. Weyl, *The Open World,* Yale University Press, 1932, Ox Bow Press, 1989.

[9] H. Weyl, *Philosophy of Mathematics and Natural Science,* Princeton University Press, 1949.

[10] G. W. Leibniz, *Philosophical Essays,* Hackett, 1989.

[11] Leibniz, *Discours de métaphysique,* Gallimard, 1995.

[12] M. Davis, *The Universal Computer: The Road from Leibniz to Turing,* Norton, 2000.

[13] J. L. Casti, *The One True Platonic Heaven,* Joseph Henry Press, 2003.

[14] E. T. Bell, *Mathematics, Queen and Servant of Science,* Tempus, 1951.

[15] G. J. Chaitin, *Algorithmic Information Theory,* Cambridge University Press, 1987.

[16] T. Ord and T. D. Kieu, "On the existence of a new family of diophantine equations for Ω," http://arxiv.org/math.NT/0301274.

[17] P. Raatikainen, Book review, *AMS Notices* **48**, Oct. 2001, pp. 992–996.

[18] H.-U. Obrist, *Interviews,* Charta, 2003.

[19] G. J. Chaitin, "Meta-mathematics and the foundations of mathematics," *EATCS Bulletin* **77**, June 2002, pp. 167–179.

[20] O. Postel-Vinay, "L'Univers est-il un calculateur?" [Is the universe a calculator?], *La Recherche* **360**, Jan. 2003, pp. 33–44.

[21] M. A. Nielsen and I. L. Chuang, *Quantum Computation and Quantum Information,* Cambridge University Press, 2000.

[22] S. Wolfram, *A New Kind of Science,* Wolfram Media, 2002.

[23] L. Smolin, *Three Roads to Quantum Gravity,* Weidenfeld and Nicolson, 2000.

[24] T. Tymoczko, *New Directions in the Philosophy of Mathematics,* Princeton University Press, 1998.

[25] J. Borwein and D. Bailey, *Mathematics by Experiment,* A. K. Peters, in press.

[26] J. Borwein and D. Bailey, *Experimentation in Mathematics,* A. K. Peters, in press.

[27] D. S. Robertson, *The New Renaissance,* Oxford University Press, 1998.

[28] D. S. Robertson, *Phase Change,* Oxford University Press, 2003.

[29] J. Maynard Smith and E. Szathmáry, *The Major Transitions in Evolution,* Oxford University Press, 1995.

[30] J. Maynard Smith and E. Szathmáry, *The Origins of Life,* Oxford University Press, 1999.

[31] C. S. Calude, *Information and Randomness,* Springer-Verlag, 2002.

[32] E. T. Bell, *Men of Mathematics,* Simon and Schuster, 1937.

[33] A. P. Coudert, *Leibniz and the Kabbalah,* Kluwer Academic, 1995.

Leibniz, randomness & the halting probability

Dedicated to Alan Turing on the 50th Anniversary of his Death

Turing's remarkable 1936 paper *On computable numbers, with an application to the Entscheidungsproblem* marks a dramatic turning point in modern mathematics. On the one hand, the computer enters center stage as a major mathematical concept. On the other hand, Turing establishes a link between mathematics and physics by talking about what a *machine* can accomplish. It is amazing how far these ideas have come in a comparatively short amount of time; a small stream has turned into a major river.

I have recently completed a small book about some of these developments, *Meta Math!* It is currently available as an e-book on my personal website, and is scheduled to be published next year. Here I will merely give a few highlights.

My story begins with Leibniz in 1686, the year before Newton published his *Principia*. Due to a snow storm, Leibniz is forced to take a break in his attempts to improve the water pumps for some important German silver mines, and writes down an outline of some of his ideas, now known to us as the *Discours de métaphysique*. Leibniz then sends a summary of the major points through a mutual friend to the famous fugitive French *philosophe* Arnauld, who is so horrified at what he reads that Leibniz never sends him nor anyone else the entire manuscript. It languishes among Leibniz's voluminous personal papers and is only discovered and published many years after Leibniz's death.

In sections V and VI of the *Discours de métaphysique,* Leibniz discusses the crucial question of how we can distinguish a world in which science applies from one in which it does not. Imagine, he says, that someone has splattered a piece of paper with ink spots determining in this manner a finite set of points on the page. Nevertheless, Leibniz observes, there will always be a mathematical equation that passes through this finite set of points. Indeed, many good ways to do this are now known. For example, what is called Lagrangian interpolation will do.

So the existence of a mathematical curve passing through a set of points cannot enable us to distinguish between points that are chosen at random and those that obey some kind of a scientific law. How then can we tell the difference? Well, says Leibniz, if the equation must be extremely complex ("fort composée") that is not a valid scientific law and the points are random ("irrégulier").

Leibniz had a million other interests and earned a living as a consultant to princes, and as far as I know after having this idea he never returned to this subject. Indeed, he was always tossing out good ideas, but rarely, with the notable exception of the infinitesimal calculus, had the time to develop them in depth.

The next person to take up this subject, as far as I know, is Hermann Weyl in his 1932 book *The Open World,* consisting of three lectures on metaphysics that Weyl gave at Yale University. In fact, I discovered Leibniz's work on complexity and randomness by reading this little book by Weyl. And Weyl points out that Leibniz's way of distinguishing between points that are random and those that follow a law by invoking the complexity of a mathematical formula is unfortunately not too well defined, since it depends on what primitive functions you are allowed to use in writing that formula and therefore varies as a function of time.

Well, the field that I invented in 1965 and which I call algorithmic information theory provides a possible solution for the problem noticed by Hermann Weyl. This theory defines a string of bits to be random, irreducible, structureless, if it is algorithmically incompressible, that is to say, if the size of the smallest computer program that produces that particular finite string of bits as output is about the same size as the output it produces.

So we have added two ideas to Leibniz's 1686 proposal. First, we measure complexity in terms of bits of information, i.e., 0s and 1s. Second, instead of mathematical equations, we use binary computer programs. Crucially, this enables us to compare the complexity of a scientific theory (the computer program) with the complexity of the data that it explains (the output of the computer program).

As Leibniz observed, for any data there is always a complicated theory, which is a computer program that is the same size as the data. But that doesn't count. It is only a real theory if there is compression, if the program is much smaller than its output, both measured in 0/1 bits. And if there can be no proper theory, then the bit string is algorithmically random or irreducible. That's how you define a random string in algorithmic information theory.

I should point out that Leibniz had the two key ideas that you need to get this modern definition of randomness, he just never made the connection. For Leibniz produced one of the first calculating machines, which he displayed at the Royal Society in London, and he was also one of the first people to appreciate base-two binary arithmetic and the fact that everything can be represented using only 0s and 1s. So, as Martin Davis argues in his book *The Universal Computer: The Road from Leibniz to Turing,* Leibniz was the first computer scientist, and he was also the first information theorist. I am sure that Leibniz would have instantly understood and appreciated the modern definition of randomness.

I should mention that A. N. Kolmogorov also proposed this definition of randomness. He and I did this independently in 1965. Kolmogorov was at the end of his career, and I was a teenager at the beginning of my own career as a mathematician. As far as I know, neither of us was aware of the Leibniz *Discours.* Let me compare Kolmogorov's work in this area with my own. I think that there are two key points to note.

Firstly, Kolmogorov never realised as I did that our original definition of randomness was incorrect. It was a good initial idea but it was technically flawed. Nine years after he and I independently proposed this definition, I realised that it was important for the computer programs that are used in the theory to be what I call "self-delimiting"; without this it is not even possible to define my Ω number that I'll discuss below. And there are other important changes that I had to make in the original definitions that Kolmogorov never realised were necessary.

Secondly, Kolmogorov thought that the key application of these ideas was to be to obtain a new, algorithmic version of probability theory. It's true, that can be done, but it's not very interesting, it's too systematic a re-reading of standard probability theory. In fact, every statement that is true with probability one, merely becomes a statement that must necessarily be true, for sure, for what are defined to be the random infinite sequences of bits. Kolmogorov never realised as I did that the really important application of these ideas was the new light that they shed on Gödel's incompleteness theorem and on Turing's halting problem.

So let me tell you about that now, and I'm sure that Turing would have loved these ideas if his premature death had not prevented him from learning about them. I'll tell you how my Ω number, which is defined to be the halting probability of a binary program whose bits are generated using independent tosses of a fair coin, shows that in a sense there is randomness in pure mathematics.

Instead of looking at individual instances of Turing's famous halting problem, you just put all possible computer programs into a bag, shake it well, pick out a program, and ask what is the probability that it will eventually halt. That's how you define the halting probability Ω, and for this to work it's important that the programs have to be self-delimiting. Otherwise the halting probability diverges to infinity instead of being a real number between zero and one like all probabilities have to be. You'll have to take my word for this; I can't explain this in detail here.

Anyway, once you do things properly you can define a halting probability Ω between zero and one. The particular value of Ω that you get depends on your choice of computer programming language, but its surprising properties don't depend on that choice.

And what is Ω's most surprising property? It's the fact that if you write Ω in binary, the bits in its base-two expansion, the bits after the binary decimal point, seem to have absolutely no mathematical structure. Even though Ω has a simple mathematical definition, its individual bits seem completely patternless. In fact, they are maximally unknowable, they have, so to speak, maximum entropy. Even though they are precisely defined once you specify the programming language, the individual bits are maximally unknowable, maximally irreducible. They seem to be mathematical facts that are true for no reason.

Why? Well, it is impossible to compress N bits of Ω into a computer program that is substantially smaller than N bits in size (so that Ω satisfies the definition of randomness of algorithmic information theory). But not only does computation fail to compress Ω, reason fails as well. No formal mathematical theory whose axioms have less than N bits of complexity can enable us to determine N bits of Ω. In other words, essentially the only way to be able to prove what the values of N bits of Ω are, is to assume what you want to prove as an axiom, which of course is cheating and doesn't really count, because you are not using reasoning at all. However, in the case of Ω, that is the best that you can ever do!

So this is an area in which mathematical truth has absolutely no structure, no structure that we will ever be able to appreciate in detail, only statistically. The best way of thinking about the bits of Ω is to say that each bit has probability $1/2$ of being zero and probability $1/2$ of being one, even though each bit is mathematically determined.

So that's where Turing's halting problem has led us, to the discovery of pure randomness in a part of mathematics. I think that Turing and Leibniz would be delighted at this remarkable turn of events.

Now I'd like to make a few comments about what I see as the philosophical implications of all of this. These are just my views, and they are quite controversial. For example, even though a recent critical review of two of my books in the *Notices of the American Mathematical Society* does not claim that there are any technical mistakes in my work, the reviewer strongly disagrees with my philosophical conclusions, and in fact he claims that my work has no philosophical implications whatsoever. So these are just my views, they are certainly not a community consensus, not at all.

My view is that Ω is a much more disagreeable instance of mathematical incompleteness than the one found by Gödel in 1931, and that it therefore forces our hand philosophically. In what way? Well, in my opinion, in a quasi-empirical direction, which is a phrase coined by Imre Lakatos when he was doing philosophy in England after leaving Hungary in 1956. In my opinion, Ω suggests that even though math and physics are different, perhaps they are not as different as most people think.

What do I mean by this? (And whether Lakatos would agree or not, I cannot say.) I think that physics enables us to compress our experimental data, and math enables us to compress the results of our computations, into scientific or mathematical theories as the case may be. And I think that neither math nor science gives absolute certainty; that is an asymptotic limit unobtainable by mortal beings. And in this connection I should mention the book (actually two books) just published by Borwein and Bailey on experimental math.

To put it bluntly, if the incompleteness phenomenon discovered by Gödel in 1931 is really serious—and I believe that Turing's work and my own work suggest that incompleteness is much more serious than people think—then perhaps mathematics should be pursued somewhat more in the spirit of experimental science rather than always demanding proofs for everything. In fact, that is what theoretical computer scientists are currently doing. Although they may not want to admit it, and refer to $P \neq NP$ as an unproved hypothesis, that community is in fact behaving as if this were a new axiom, the way that physicists would.

At any rate, that's the way things seem to me. Perhaps by the time we reach the centenary of Turing's death this quasi-empirical view will have made some headway, or perhaps instead these foreign ideas will be utterly rejected by the immune system of the math community. For now they certainly are rejected. But the past fifty years have brought us many surprises, and I expect that the next fifty years will too, a great many indeed.

References

1. Chapter on Leibniz in E. T. Bell, *Men of Mathematics,* Simon & Schuster, 1937.

2. R. C. Sleigh, Jr., *Leibniz and Arnauld: A Commentary on Their Correspondence,* Yale University Press, 1990.

3. H. Weyl, *The Open World: Three Lectures on the Metaphysical Implications of Science,* Yale University Press, 1932, Ox Bow Press, 1994.

4. First article by Lakatos in T. Tymoczko, *New Directions in the Philosophy of Mathematics,* Princeton University Press, 1998.

5. M. Davis, *The Universal Computer: The Road from Leibniz to Turing,* Norton, 2000.

6. Play on Newton vs. Leibniz in C. Djerassi, D. Pinner, *Newton's Darkness: Two Dramatic Views,* Imperial College Press, 2003.

7. J. Borwein, D. Bailey, *Mathematics by Experiment: Plausible Reasoning in the 21st Century,* A. K. Peters, 2004.

8. J. Borwein, D. Bailey, R. Girgensohn, *Experimentation in Mathematics: Computational Paths to Discovery,* A. K. Peters, 2004.

9. G. Chaitin, *Meta Math! The Quest for Omega,* to be published by Pantheon Books in 2005.

Complexity & Leibniz

*Inaugural **Académie Internationale de Philosophie des Sciences** lecture by Gregory Chaitin, Tenerife, September 2005.*

> *Cum Deus calculat, fit mundus.*
> As God calculates, so the world is made.
> —***Leibniz***

I am a mathematician and my field is algorithmic information theory (AIT). AIT deals with program-size complexity or algorithmic information content, which I regard more or less as the **complexity of ideas**. I think that this has much greater philosophical significance than the much more popular complexity concepts based on time or other measures of **computational effort or work**.

Thank you very much for making me a member of the *Académie Internationale de Philosophie des Sciences*. And thanks for squeezing me into the program and making space for me to give a short talk. Thank you very much for asking me to give a talk even though I was not scheduled to be a speaker at this meeting.

I've brought with me, hot off the press, a copy of my new book *Meta Math!*. This book has been 40 years in the making. I've been working on these questions for that long. In this brief talk, I'll merely touch on topics that are developed at much greater length in my book.

To start the ball rolling, let's consider physics versus biology. Is mathematics more like physics or is it more like biology? Well, in physics we have **simple** equations, whereas biology is the domain of **complexity**. So normally people think that math is much closer to physics than it is to biology.

After all, mathematics and physics have co-evolved, and not much mathematics is used in biology. However, as I'll explain in this talk, mathematics contains **infinite complexity** and is therefore, in a fundamental sense, much closer to biology than it is to physics!

How does AIT manage to show this surprising and unexpected connection between mathematics and biology? AIT is at this point in time a fully developed elegant mathematical theory of program-size complexity. But for the purposes of this discussion, we, as philosophers of science, do not really need to know the mathematical details of AIT. Instead it suffices to understand the basic concepts, which amazingly enough can be traced back to Leibniz. Here are three texts by Leibniz that caught my eye as a mathematician. They are, as far as I'm aware, his key texts on the concept of **complexity**:

1. *Discours de métaphysique,* Sections 5–6:
   As Hermann Weyl put it, if an **arbitrarily complicated** law is permitted, then the concept of "law" becomes vacuous, because there is always a law!

2. *Principles of Nature and Grace,* Section 7:
   Why is there something rather than nothing? For nothing is **simpler** and easier than something! (This is a fascinating question, but it has nothing to do with AIT, which is a mathematical theory, it has more to do with physics and cosmology.)

3. *The Monadology,* Sections 33–35:
   Proof consists of reducing **complicated** assertions to **simpler** ones until assertions are reached that are self-evident or axioms. And going beyond Leibniz, let me ask you to ponder what if this is impossible, what if we have complicated **irreducible** truths?!

Having used Leibniz as an introduction, let me now leap into the heart of AIT. One of the central topics in AIT is a number that I've discovered that I like to call Ω. Briefly, Ω is the halting probability of a computer, it's equal to 2 raised to the power $-K$ summed over the size in bits K of every program that ever halts:

$$0 < \Omega = \sum_{p \text{ halts}} 2^{-(\text{size in bits of } p)} < 1.$$

Ω is important because it's an oracle for Turing's halting problem, it's the most compact, the most concise way of summarizing all the possible answers to questions asking

"Does a particular computer program p ever halt?,"

as originally discussed by Turing in 1936. Ω is irredundant, the infinite stream of base-two bits in its binary expansion are **irreducible** mathematical facts. In other words, whether each bit in the base-two binary expansion of Ω is a 0 or a 1 is a mathematical fact that is true for no reason, no reason simpler than just directly knowing the bits themselves. More precisely:

You need an N-bit mathematical theory—one with N bits of axioms—in order to be able to determine N bits of Ω.

Please note that this information-theoretic limitative meta-theorem contradicts Leibniz's **principle of sufficient of reason**, which says that if something is true then it has to be true for a reason. (Of course this applies only to **necessary** not to **contingent** truths.) Those reasons as Leibniz points out in *The Monadology* would necessarily have to be *simpler* than the bits of Ω in order to be able to count as the reasons determining their individual 0/1 values. But in the case of Ω, which is irreducible, no simpler reasons are possible. In other words, the bits of Ω are **logically** as well as **computationally** irreducible, that is why they refute the principle of sufficient reason. Essentially the only way to establish what these bits are is to add that information directly to your mathematical theory as a new axiom. But anything can be established by adding it as a new axiom. That's not using reasoning, that's not much of a proof, it's a new assumption.

Furthermore, *in toto* the bits of Ω are infinitely complex, which establishes the promised link between mathematics and biology.

To conclude, I would like to thank you all again for making me a member of this Academy. I know I've rushed through this material very, very quickly. But if you want to know more about all of this, please take a look at my new book. In fact, it's actually my *système du monde*, it's my attempt to formulate a complete speculative metaphysics. *Meta Math!* is a serious book in spite of the frivolous-sounding title. For example, let me mention three important topics in my book that I haven't had time to discuss here:

1. My "quasi-empirical" view of mathematics.

2. The ontological status of real numbers, which in my opinion are unreal.

3. The ontological status of discrete binary information, which in my opinion is real even though it may not have a material basis (no physical implementation or recording technology).

Thank you.

References

- G. J. Chaitin, *Meta Math!*, 2005.
- G. W. Leibniz, *Discours de métaphysique*, 1686.
- G. W. Leibniz, *Principles of Nature and Grace*, 1714.
- G. W. Leibniz, *The Monadology*, 1714.
- H. Weyl, *The Open World*, 1932.

The limits of reason

Ideas on complexity and randomness originally suggested by Gottfried W. Leibniz in 1686, combined with modern information theory, imply that there can never be a "theory of everything" for all of mathematics.

In 1956 *Scientific American* published an article by Ernest Nagel and James R. Newman entitled "Gödel's Proof." Two years later the writers published a book with the same title—a wonderful work that is still in print. I was a child, not even a teenager, and I was obsessed by this little book. I remember the thrill of discovering it in the New York Public Library. I used to carry it around with me and try to explain it to other children.

It fascinated me because Kurt Gödel used mathematics to show that mathematics itself has limitations. Gödel refuted the position of David Hilbert, who about a century ago declared that there was a theory of everything for math, a finite set of principles from which one could mindlessly deduce all mathematical truths by tediously following the rules of symbolic mathematical logic. But Gödel demonstrated that mathematics contains true statements that cannot be proved that way. His result is based on two self-referential paradoxes: "This statement is false" and "This statement is unprovable." (For more on Gödel's incompleteness theorem, see Box 1.)

My attempt to understand Gödel's proof took over my life, and now half a century later I have published a little book of my own. In some respects, it is my own version of Nagel and Newman's book, but it does not focus on Gödel's proof. The only things the two books have in common are their small size and their goal of critiquing mathematical methods.

Unlike Gödel's approach, mine is based on measuring information and showing that some mathematical facts cannot be compressed into a theory because they are too complicated. This new approach suggests that what Gödel discovered was just the tip of the iceberg: an infinite number of true mathematical theorems exist that cannot be proved from any finite system of axioms.

Complexity and Scientific Laws

My story begins in 1686 with Gottfried W. Leibniz's philosophical essay *Discours de métaphysique* (*Discourse on Metaphysics*), in which he discusses how one can distinguish between facts that can be described by some law and those that are lawless, irregular facts. Leibniz's very simple and profound idea appears in section VI of the *Discours,* in which he essentially states that a theory has to be simpler than the data it explains, otherwise it does not explain anything. The concept of a law becomes vacuous if arbitrarily high mathematical complexity is permitted, because then one can always construct a law no matter how random and patternless the data really are. Conversely, if the only law that describes some data is an extremely complicated one, then the data are actually lawless.

Today the notions of complexity and simplicity are put in precise quantitative terms by a modern branch of mathematics called algorithmic information theory. Ordinary information theory quantifies information by asking how many bits are needed to encode the information. For example, it takes one bit to encode a single yes/no answer. Algorithmic information, in contrast, is defined by asking what size computer program is necessary to generate the data. The minimum number of bits—what size string of zeros and ones—needed to store the program is called the algorithmic information content of the data. Thus, the infinite sequence of numbers 1, 2, 3, ... has very little algorithmic information; a very short computer program can generate all those numbers. It does not matter how long the program must take to do the computation or how much memory it must use—just the length of the program in bits counts. (I gloss over the question of what programming language is used to write the program—for a rigorous definition, the language would have to be specified precisely. Different programming languages would result in somewhat different values of algorithmic information content.)

To take another example, the number π, 3.14159..., also has only a little algorithmic information content, because a relatively short algorithm can be programmed into a computer to compute digit after digit. In contrast, a random number with a mere million digits, say $1.341285\ldots 64$, has a much larger amount of algorithmic information. Because the number lacks a defining pattern, the shortest program for outputting it will be about as long as the number itself:

```
Begin
Print "1.341285...64"
End
```

(All the digits represented by the ellipsis are included in the program.) No smaller program can calculate that sequence of digits. In other words, such digit streams are incompressible, they have no redundancy; the best that one can do is transmit them directly. They are called irreducible or algorithmically random.

How do such ideas relate to scientific laws and facts? The basic insight is a software view of science: a scientific theory is like a computer program that predicts our observations, the experimental data. Two fundamental principles inform this viewpoint. First, as William of Occam noted, given two theories that explain the data, the simpler theory is to be preferred (Occam's razor). That is, the smallest program that calculates the observations is the best theory. Second is Leibniz's insight, cast in modern terms—if a theory is the same size in bits as the data it explains, then it is worthless, because even the most random of data has a theory of that size. A useful theory is a compression of the data; comprehension is compression. You compress things into computer programs, into concise algorithmic descriptions. The simpler the theory, the better you understand something.

Sufficient Reason

Despite living 250 years before the invention of the computer program, Leibniz came very close to the modern idea of algorithmic information. He had all the key elements. He just never connected them. He knew that everything can be represented with binary information, he built one of the first calculating machines, he appreciated the power of computation, and he discussed complexity and randomness.

If Leibniz had put all this together, he might have questioned one of the key pillars of his philosophy, namely, the principle of sufficient reason—that everything happens for a reason. Furthermore, if something is true, it must be true for a reason. That may be hard to believe sometimes, in the confusion and chaos of daily life, in the contingent ebb and flow of human history. But even if we cannot always see a reason (perhaps because the chain of reasoning is long and subtle), Leibniz asserted, God can see the reason. It is there! In that, he agreed with the ancient Greeks, who originated the idea.

Mathematicians certainly believe in reason and in Leibniz's principle of sufficient reason, because they always try to prove everything. No matter how much evidence there is for a theorem, such as millions of demonstrated examples, mathematicians demand a proof of the general case. Nothing less will satisfy them.

And here is where the concept of algorithmic information can make its surprising contribution to the philosophical discussion of the origins and limits of knowledge. It reveals that certain mathematical facts are true for no reason, a discovery that flies in the face of the principle of sufficient reason.

Indeed, as I will show later, it turns out that an infinite number of mathematical facts are irreducible, which means no theory explains why they are true. These facts are not just computationally irreducible, they are logically irreducible. The only way to "prove" such facts is to assume them directly as new axioms, without using reasoning at all.

The concept of an "axiom" is closely related to the idea of logical irreducibility. Axioms are mathematical facts that we take as self-evident and do not try to prove from simpler principles. All formal mathematical theories start with axioms and then deduce the consequences of these axioms, which are called its theorems. That is how Euclid did things in Alexandria two millennia ago, and his treatise on geometry is the classical model for mathematical exposition.

In ancient Greece, if you wanted to convince your fellow citizens to vote with you on some issue, you had to reason with them—which I guess is how the Greeks came up with the idea that in mathematics you have to prove things rather than just discover them experimentally. In contrast, previous cultures in Mesopotamia and Egypt apparently relied on experiment. Using reason has certainly been an extremely fruitful approach, leading to modern mathematics and mathematical physics and all that goes with them, including the technology for building that highly logical and mathematical machine, the computer.

So am I saying that this approach that science and mathematics has been following for more than two millennia crashes and burns? Yes, in a sense I am. My counterexample illustrating the limited power of logic and reason, my source of an infinite stream of unprovable mathematical facts, is the number that I call Ω.

The Number Omega

The first step on the road to Ω came in a famous paper published precisely 250 years after Leibniz's essay. In a 1936 issue of the *Proceedings of the London Mathematical Society,* Alan M. Turing began the computer age by presenting a mathematical model of a simple, general-purpose, programmable digital computer. He then asked, Can we determine whether or not a computer program will ever halt? This is Turing's famous halting problem.

Of course, by running a program you can eventually discover that it halts, if it halts. The problem, and it is an extremely fundamental one, is to decide when to give up on a program that does not halt. A great many special cases can be solved, but Turing showed that a general solution is impossible. No algorithm, no mathematical theory, can ever tell us which programs will halt and which will not. (For a modern proof of Turing's thesis, see Box 2.) By the way, when I say "program," in modern terms I mean the concatenation of the computer program and the data to be read in by the program.

The next step on the path to the number Ω is to consider the ensemble of all possible programs. Does a program chosen at random ever halt? The probability of having that happen is my Ω number. First, I must specify how to pick a program at random. A program is simply a series of bits, so flip a coin to determine the value of each bit. How many bits long should the program be? Keep flipping the coin so long as the computer is asking for another bit of input. Ω is just the probability that the machine will eventually come to a halt when supplied with a stream of random bits in this fashion. (The precise numerical value of Ω depends on the choice of computer programming language, but Ω's surprising properties are not affected by this choice. And once you have chosen a language, Ω has a definite value, just like π or the number 3.)

Being a probability, Ω has to be greater than 0 and less than 1, because some programs halt and some do not. Imagine writing Ω out in binary. You would get something like 0.1110100... These bits after the decimal point form an irreducible stream. They are our irreducible mathematical facts (each fact being whether the bit is a 0 or a 1).

Ω can be defined as an infinite sum, and each N-bit program that halts contributes precisely $1/2^N$ to the sum [see Box 3]. In other words, each N-bit program that halts adds a 1 to the Nth bit in the binary expansion of Ω. Add up all the bits for all programs that halt, and you would get the precise value of Ω. This description may make it sound like you can calculate Ω accurately, just as if it were $\sqrt{2}$ or the number π. Not so—Ω is perfectly well defined and it is a specific number, but it is impossible to compute in its entirety.

We can be sure that Ω cannot be computed because knowing Ω would let us solve Turing's halting problem, but we know that this problem is unsolvable. More specifically, knowing the first N bits of Ω would enable you to decide whether or not each program up to N bits in size ever halts [see Box 4]. From this it follows that you need at least an N-bit program to calculate N bits of Ω.

Note that I am not saying that it is impossible to compute some digits of Ω. For example, if we knew that computer programs 0, 10 and 110 all halt, then we would know that the first digits of Ω were 0.111. The point is that the first N digits of Ω cannot be computed using a program significantly shorter than N bits long.

Most important, Ω supplies us with an infinite number of these irreducible bits. Given any finite program, no matter how many billions of bits long, we have an infinite number of bits that the program cannot compute. Given any finite set of axioms, we have an infinite number of truths that are unprovable in that system.

Because Ω is irreducible, we can immediately conclude that a theory of everything for all of mathematics cannot exist. An infinite number of bits of Ω constitute mathematical facts (whether each bit is a 0 or a 1) that cannot be derived from any principles simpler than the string of bits itself. Mathematics therefore has infinite complexity, whereas any individual theory of everything would have only finite complexity and could not capture all the richness of the full world of mathematical truth.

This conclusion does not mean that proofs are no good, and I am certainly not against reason. Just because some things are irreducible does not mean we should give up using reasoning. Irreducible principles—axioms—have always been a part of mathematics. Ω just shows that a lot more of them are out there than people suspected.

So perhaps mathematicians should not try to prove everything. Sometimes they should just add new axioms. That is what you have got to do if you are faced with irreducible facts. The problem is realizing that they are irreducible! In a way, saying something is irreducible is giving up, saying that it cannot ever be proved. Mathematicians would rather die than do that, in sharp contrast with their physicist colleagues, who are happy to be pragmatic and to use plausible reasoning instead of rigorous proof. Physicists are willing to add new principles, new scientific laws, to understand new domains of experience. This raises what I think is an extremely interesting question: Is mathematics like physics?

Mathematics and Physics

The traditional view is that mathematics and physics are quite different. Physics describes the universe and depends on experiment and observation. The particular laws that govern our universe—whether Newton's laws of motion or the Standard Model of particle physics—must be determined empirically and then asserted like axioms that cannot be logically proved, merely verified.

Mathematics, in contrast, is somehow independent of the universe. Results and theorems, such as the properties of the integers and real numbers, do not depend in any way on the particular nature of reality in which we find ourselves. Mathematical truths would be true in any universe.

Yet both fields are similar. In physics, and indeed in science generally, scientists compress their experimental observations into scientific laws. They then show how their observations can be deduced from these laws. In mathematics, too, something like this happens—mathematicians compress their computational experiments into mathematical axioms, and they then show how to deduce theorems from these axioms.

If Hilbert had been right, mathematics would be a closed system, without room for new ideas. There would be a static, closed theory of everything for all of mathematics, and this would be like a dictatorship. In fact, for mathematics to progress you actually need new ideas and plenty of room for creativity. It does not suffice to grind away, mechanically deducing all the possible consequences of a fixed number of basic principles. I much prefer an open system. I do not like rigid, authoritarian ways of thinking.

Another person who thought mathematics is like physics was Imre Lakatos, who left Hungary in 1956 and later worked on philosophy of science in England. There Lakatos came up with a great word, "quasi-empirical," which means that even though there are no true experiments that can be carried out in mathematics, something similar does take place. For example, the Goldbach conjecture states that any even number greater than 2 can be expressed as the sum of two prime numbers. This conjecture was arrived at experimentally, by noting empirically that it was true for every even number that anyone cared to examine. The conjecture has not yet been proved, but it has been verified up to 10^{14}.

I think that mathematics is quasi-empirical. In other words, I feel that mathematics is different from physics (which is truly empirical) but perhaps not as different as most people think.

I have lived in the worlds of both mathematics and physics, and I never thought there was such a big difference between these two fields. It is a matter of degree, of emphasis, not an absolute difference. After all, mathematics and physics coevolved. Mathematicians should not isolate themselves. They should not cut themselves off from rich sources of new ideas.

New Mathematical Axioms

The idea of choosing to add more axioms is not an alien one to mathematics. A well-known example is the parallel postulate in Euclidean geometry: given a line and a point not on the line, there is exactly one line that can be drawn through the point that never intersects the original line. For centuries geometers wondered whether that result could be proved using the rest of Euclid's axioms. It could not. Finally, mathematicians realized that they could substitute different axioms in place of the Euclidean version, thereby producing the non-Euclidean geometries of curved spaces, such as the surface of a sphere or of a saddle.

Other examples are the law of the excluded middle in logic and the axiom of choice in set theory. Most mathematicians are happy to make use of those axioms in their proofs, although others do not, exploring instead so-called intuitionist logic or constructivist mathematics. Mathematics is not a single monolithic structure of absolute truth!

Another very interesting axiom may be the "**P** $\neq$ **NP**" conjecture. **P** and **NP** are names for classes of problems. An **NP** problem is one for which a proposed solution can be verified quickly. For example, for the problem "find the factors of 8,633," one can quickly verify the proposed solution "97 and 89" by multiplying those two numbers. (There is a technical definition of "quickly," but those details are not important here.) A **P** problem is one that can be solved quickly even without being given the solution. The question is—and no one knows the answer—can every **NP** problem be solved quickly? (Is there a quick way to find the factors of 8,633?) That is, is the class **P** the same as the class **NP**? This problem is one of the Clay Millennium Prize Problems for which a reward of $1 million is on offer.

Computer scientists widely believe that **P** $\neq$ **NP**, but no proof is known. One could say that a lot of quasi-empirical evidence points to **P** not being equal to **NP**. Should **P** $\neq$ **NP** be adopted as an axiom, then? In effect, this is what the computer science community has done. Closely related to this issue is the security of certain cryptographic systems used throughout the world. The systems are believed to be invulnerable to being cracked, but no one can prove it.

Experimental Mathematics

Another area of similarity between mathematics and physics is experimental mathematics: the discovery of new mathematical results by looking at many examples using a computer. Whereas this approach is not as persuasive as a short proof, it can be more convincing than a long and extremely complicated proof, and for some purposes it is quite sufficient.

In the past, this approach was defended with great vigor by George Pólya and Lakatos, believers in heuristic reasoning and in the quasi-empirical nature of mathematics. This methodology is also practiced and justified in Stephen Wolfram's *A New Kind of Science* (2002).

Extensive computer calculations can be extremely persuasive, but do they render proof unnecessary? Yes and no. In fact, they provide a different kind of evidence. In important situations, I would argue that both kinds of evidence are required, as proofs may be flawed, and conversely computer searches may have the bad luck to stop just before encountering a counterexample that disproves the conjectured result.

All these issues are intriguing but far from resolved. It is now 2006, 50 years after this magazine published its article on Gödel's proof, and we still do not know how serious incompleteness is. We do not know if incompleteness is telling us that mathematics should be done somewhat differently. Maybe 50 years from now we will know the answer.

Overview/Irreducible Complexity

- Kurt Gödel demonstrated that mathematics is necessarily incomplete, containing true statements that cannot be formally proved. A remarkable number known as Ω reveals even greater incompleteness by providing an infinite number of theorems that cannot be proved by any finite system of axioms. A "theory of everything" for mathematics is therefore impossible.

- Ω is perfectly well defined [see Box 3] and has a definite value, yet it cannot be computed by any finite computer program.

- Ω's properties suggest that mathematicians should be more willing to postulate new axioms, similar to the way that physicists must evaluate experimental results and assert basic laws that cannot be proved logically.

- The results related to Ω are grounded in the concept of algorithmic information. Gottfried W. Leibniz anticipated many of the features of algorithmic information theory more than 300 years ago.

Box 1. What Is Gödel's Proof?

Kurt Gödel's incompleteness theorem demonstrates that mathematics contains true statements that cannot be proved. His proof achieves this by constructing paradoxical mathematical statements. To see how the proof works, begin by considering the liar's paradox: "This statement is false." This statement is true if and only if it is false, and therefore it is neither true nor false.

Now let's consider "This statement is unprovable." If it is provable, then we are proving a falsehood, which is extremely unpleasant and is generally assumed to be impossible. The only alternative left is that this statement is unprovable. Therefore, it is in fact both true and unprovable. Our system of reasoning is incomplete, because some truths are unprovable.

Gödel's proof assigns to each possible mathematical statement a so-called Gödel number. These numbers provide a way to talk about properties of the statements by talking about the numerical properties of very large integers.

Gödel uses his numbers to construct self-referential statements analogous to the plain English paradox "This statement is unprovable."

Strictly speaking, his proof does not show that mathematics is incomplete. More precisely, it shows that individual formal axiomatic mathematical theories fail to prove the true numerical statement "This statement is unprovable." These theories therefore cannot be "theories of everything" for mathematics.

The key question left unanswered by Gödel: Is this an isolated phenomenon, or are there many important mathematical truths that are unprovable?—*G.C.*

Box 2. Why Is Turing's Halting Problem Unsolvable?

A key step in showing that incompleteness is natural and pervasive was taken by Alan M. Turing in 1936, when he demonstrated that there can be no general procedure to decide if a self-contained computer program will eventually halt.

To demonstrate this result, let us assume the opposite of what we want to prove is true. Namely, assume that there is a general procedure H that can decide whether any given computer program will halt. From this assumption we shall derive a contradiction. This is what is called a *reductio ad absurdum* proof.

So assuming the existence of H, we can construct the following program P that uses H as a subroutine. The program P knows its own size in bits (N)—there is certainly room in P for it to contain the number N—and then using H, which P contains, P takes a look at all programs up to 100 times N bits in size to see which halt and which do not. Then P runs all the ones that halt to determine the output that they produce. This will be precisely the set of all digital objects with complexity up to 100 times N. Finally, our program P outputs the smallest positive integer not in this set, and then P itself halts.

So P halts, P's size is N bits, and P's output is an integer that cannot be produced by a program whose size is less than or equal to 100 times N bits. But P has just produced this integer as its output, and it is much too small to be able to do this, because P's size is only N bits, which is much less than 100 times N. Contradiction! Therefore, a general procedure H for deciding whether or not programs ever halt cannot exist, for if it did then we could actually construct this paradoxical program P using H.

Finally, Turing points out that if there were a theory of everything that always enables you to prove that an individual program halts or to prove that it never does, whichever is the case, then by systematically running through all possible proofs you could eventually decide whether individual programs ever halt. In other words, we could use this theory to construct H, which we have just shown cannot exist. Therefore there is no theory of everything for the halting problem.

Similar reasoning shows that no program that is substantially shorter than N bits long can solve the Turing halting problem for all programs up to N bits long.—*G.C.*

Box 3. How Omega Is Defined

To see how the value of the number Ω is defined, look at a simplified example. Suppose that the computer we are dealing with has only three programs that halt, and they are the bit strings 110, 11100 and 11110. These programs are, respectively, 3, 5 and 5 bits in size. If we are choosing programs at random by flipping a coin for each bit, the probability of getting each of them by chance is precisely $1/2^3$, $1/2^5$ and $1/2^5$, because each particular bit has probability $1/2$. So the value of Ω (the halting probability) for this particular computer is given by the equation:

$$
\begin{aligned}
\Omega &= 1/2^3 + 1/2^5 + 1/2^5 \\
&= .001 + .00001 + .00001 \\
&= .00110
\end{aligned}
$$

This binary number is the probability of getting one of the three halting programs by chance. Thus, it is the probability that our computer will halt. Note that because program 110 halts we do not consider any programs that start with 110 and are larger than three bits—for example, we do not consider 1100 or 1101. That is, we do not add terms of .0001 to the sum for each of those programs. We regard all the longer programs, 1100 and so on, as being included in the halting of 110. Another way of saying this is that the programs are self-delimiting; when they halt, they stop asking for more bits.—*G.C.*

Box 4. Why Is Omega Incompressible?

I wish to demonstrate that Ω is incompressible—that one cannot use a program substantially shorter than N bits long to compute the first N bits of Ω. The demonstration will involve a careful combination of facts about Ω and the Turing halting problem that it is so intimately related to. Specifically, I will use the fact that the halting problem for programs up to length N bits cannot be solved by a program that is itself shorter than N bits (see Box 2).

My strategy for demonstrating that Ω is incompressible is to show that having the first N bits of Ω would tell me how to solve the Turing halting problem for programs up to length N bits. It follows from that conclusion that no program shorter than N bits can compute the first N bits of Ω. (If such a program existed, I could use it to compute the first N bits of Ω and then use those bits to solve Turing's problem up to N bits—a task that is impossible for such a short program.)

Now let us see how knowing N bits of Ω would enable me to solve the halting problem—to determine which programs halt—for all programs up to N bits in size. Do this by performing a computation in stages. Use the integer K to label which stage we are at: $K = 1, 2, 3, \ldots$

At stage K, run every program up to K bits in size for K seconds. Then compute a halting probability, which we will call Ω_K, based on all the programs that halt by stage K. Ω_K will be less than Ω because it is based on only a subset of all the programs that halt eventually, whereas Ω is based on *all* such programs.

As K increases, the value of Ω_K will get closer and closer to the actual value of Ω. As it gets closer to Ω's actual value, more and more of Ω_K's first bits will be correct—that is, the same as the corresponding bits of Ω.

And as soon as the first N bits are correct, you know that you have encountered every program up to N bits in size that will ever halt. (If there were another such N-bit program, at some later-stage K that program would halt, which would increase the value of Ω_K to be greater than Ω, which is impossible.)

So we can use the first N bits of Ω to solve the halting problem for all programs up to N bits in size. Now suppose we could compute the first N bits of Ω with a program substantially shorter than N bits long. We could then combine that program with the one for carrying out the Ω_K algorithm, to produce a program shorter than N bits that solves the Turing halting problem up to programs of length N bits.

But, as stated up front, we know that no such program exists. Consequently, the first N bits of Ω must require a program that is almost N bits long to compute them. That is good enough to call Ω incompressible or irreducible. (A compression from N bits to almost N bits is not significant for large N.)—*G.C.*

THE AUTHOR

Gregory Chaitin is a researcher at the IBM Thomas J. Watson Research Center. He is also honorary professor at the University of Buenos Aires and visiting professor at the University of Auckland. He is co-founder, with Andrei N. Kolmogorov, of the field of algorithmic information theory. His nine books include the nontechnical works *Conversations with a Mathematician* (2002) and *Meta Math!* (2005). When he is not thinking about the foundations of mathematics, he enjoys hiking and snowshoeing in the mountains.

MORE TO EXPLORE

For a chapter on Leibniz, see **Men of Mathematics.** E. T. Bell. Reissue. Touchstone, 1986.

For more on a quasi-empirical view of math, see **New Directions in the Philosophy of Mathematics.** Edited by Thomas Tymoczko. Princeton University Press, 1998.

Gödel's Proof. Revised edition. E. Nagel, J. R. Newman and D. R. Hofstadter. New York University Press, 2002.

Mathematics by Experiment: Plausible Reasoning in the 21st Century. J. Borwein and D. Bailey. A. K. Peters, 2004.

For Gödel as a philosopher and the Gödel-Leibniz connection, see **Incompleteness: The Proof and Paradox of Kurt Gödel.** Rebecca Goldstein. W. W. Norton, 2005.

Meta Math!: The Quest for Omega. Gregory Chaitin. Pantheon Books, 2005.

Short biographies of mathematicians can be found at `www-history.mcs.st-andrews.ac.uk/BiogIndex.html`

Gregory Chaitin's home page is `www.cs.auckland.ac.nz/~chaitin/`

How real are real numbers?

We discuss mathematical and physical arguments against continuity and in favor of discreteness, with particular emphasis on the ideas of Emile Borel (1871–1956).

1. Introduction

Experimental physicists know how difficult accurate measurements are. No physical quantity has ever been measured with more than 15 or so digits of accuracy. Mathematicians, however, freely fantasize with infinite-precision real numbers. Nevertheless within pure math the notion of a real number is extremely problematic.

We'll compare and contrast two parallel historical episodes:

1. the diagonal and probabilistic proofs that reals are uncountable, and

2. the diagonal and probabilistic proofs that there are uncomputable reals.

Both case histories open chasms beneath the feet of mathematicians. In the first case these are the famous Jules Richard paradox (1905), Emile Borel's know-it-all real (1927), and the fact that most reals are unnameable, which was the subject of [Borel, 1952], his last book, published when Borel was 81 years old [James, 2002].

In the second case the frightening features are the unsolvability of the halting problem (Turing, 1936), the fact that most reals are uncomputable, and last but not least, the halting probability Ω, which is irreducibly complex (algorithmically random), maximally unknowable, and dramatically illustrates the limits of reason [Chaitin, 2005].

In addition to this mathematical soul-searching regarding real numbers, some physicists are beginning to suspect that the physical universe is actually discrete [Smolin, 2000] and perhaps even a giant computer [Fredkin, 2004, Wolfram, 2002]. It will be interesting to see how far this so-called "digital philosophy," "digital physics" viewpoint can be taken.

Nota bene: To simplify matters, throughout this paper we restrict ourselves to reals in the interval between 0 and 1. We can therefore identify a real number with the infinite sequence of digits or bits after its decimal or binary point.

2. Reactions to Cantor's Theory of Sets: The Trauma of the Paradoxes of Set Theory

Cantor's theory of infinite sets, developed in the late 1800's, was a decisive advance for mathematics, but it provoked raging controversies and abounded in paradox. One of the first books by the distinguished French mathematician Emile Borel (1871–1956)[1] was his *Leçons sur la Théorie des Fonctions* [Borel, 1950], originally published in 1898, and subtitled *Principes de la théorie des ensembles en vue des applications à la théorie des fonctions.*

This was one of the first books promoting Cantor's theory of sets (*ensembles*), but Borel had serious reservations about certain aspects of Cantor's theory, which Borel kept adding to later editions of his book as new appendices. The final version of Borel's book, which was published by Gauthier-Villars in 1950, has been kept in print by Gabay. That's the one that I have, and this book is a treasure trove of interesting mathematical, philosophical and historical material.

One of Cantor's crucial ideas is the distinction between the denumerable or countable infinite sets, such as the positive integers or the rational numbers, and the much larger nondenumerable or uncountable infinite sets, such as the real numbers or the points in the plane or in space. Borel had constructivist leanings, and as we shall see he felt comfortable with denumerable sets, but very uncomfortable with nondenumerable ones. And one of Cantor's key results that is discussed by Borel is Cantor's proof that the set of reals is nondenumerable, i.e., cannot be placed in a one-to-one correspondence with the positive integers. I'll prove this now in two different ways.

---

[1]For a biography of Borel, see [James, 2002].

2.1. Cantor's diagonal argument: Reals are uncountable/nondenumerable

Cantor's proof of this is a *reductio ad absurdum*.

Suppose on the contrary that we have managed to list all the reals, with a first real, a second real, etc. Let $d(i, j)$ be the jth digit after the decimal point of the ith real in the list. Consider the real r between 0 and 1 whose kth digit is defined to be 4 if $d(k, k) = 3$, and 3 otherwise. In other words, we form r by taking all the decimal digits on the **diagonal** of the list of all reals, and then changing each of these diagonal digits.

The real r differs from the ith real in this presumably complete list of all reals, because their ith digits are different. Therefore this list cannot be complete, and the set of reals is uncountable. *Q.E.D.*

Nota bene: The most delicate point in this proof is to avoid having r end in an infinity of 0's or an infinity of 9's, to make sure that having its kth digit differ from the kth digit of the kth real in the list suffices to guarantee that r is not equal to the kth real in the list. This is how we get around the fact that some reals can have more than one decimal representation.

2.2. Alternate proof: Any countable/denumerable set of reals has measure zero

Now here is a radically different proof that the reals are uncountable. This proof, which I learned in [Courant & Robbins, 1947], was perhaps or at least **could have been** originally discovered by Borel, because it uses the mathematical notion of *measure*, which was invented by Borel and later perfected by his Ecole Normale Supérieure student Lebesgue, who now usually gets all the credit.

Measure theory and probability theory are really one and the same—it's just different names for the same concepts. And Borel was interested in both the technical mathematical aspects and in the many important practical applications, which Borel discussed in many of his books.

So let's suppose we are given a real $\epsilon > 0$, which we shall later make arbitrarily small. Consider again that supposedly complete enumeration of all the reals, a first one, a second one, etc. Cover each real with an interval, and take the interval for covering the ith real in the list to be of length $\epsilon/2^i$.

The total length of all the covering intervals is therefore

$$\frac{\epsilon}{2} + \frac{\epsilon}{4} + \cdots \frac{\epsilon}{2^i} + \cdots \quad = \quad \epsilon,$$

which we can make as small as we wish.

In other words, any countable set of reals has *measure zero* and is a so-called *null set,* i.e., has zero probability and is an infinitesimal subset of the set of all reals. *Q.E.D.*

We have now seen the two fundamentally different ways of showing that the reals are infinitely more numerous than the positive integers, i.e., that the set of all reals is a higher-order infinity than the set of all positive integers.

So far, so good! But now, let's show what a minefield this is.

2.3. Richard's paradox: Diagonalize over all nameable reals ⟶ a nameable, unnameable real

The problem is that the set of reals is uncountable, but the set of all possible texts in English or French is countable, and so is the set of all possible mathematical definitions or the set of all possible mathematical questions, since these also have to be formulated within a language, yielding at most a denumerable infinity of possibilities. So there are too many reals, and not enough texts.

The first person to notice this difficulty was Jules Richard in 1905, and the manner in which he formulated the problem is now called Richard's paradox.

Here is how it goes. Since all possible texts in French (Richard was French) can be listed or enumerated, a first text, a second one, etc.,[2] you can diagonalize over all the reals that can be defined or named in French and produce a real number that cannot be defined and is therefore unnameable. However, we've just indicated how to define it or name it!

In other words, Richard's paradoxical real differs from every real that is definable in French, but nevertheless can itself be defined in French by specifying in detail how to apply Cantor's diagonal method to the list of all possible mathematical definitions for individual real numbers in French!

How very embarrassing! Here is a real number that is simultaneously nameable yet at the same time it cannot be named using any text in French.

---

[2]List all possible texts in size order, and within texts that are the same size, in alphabetical order.

2.4. Borel's know-it-all number

The idea of being able to list or enumerate all possible texts in a language is an extremely powerful one, and it was exploited by Borel in 1927 [Tasić, 2001, Borel, 1950] in order to define a real number that can answer every possible yes/no question!

You simply write this real in binary, and use the nth bit of its binary expansion to answer the nth question in French.

Borel speaks about this real number ironically. He insinuates that it's illegitimate, unnatural, artificial, and that it's an "unreal" real number, one that there is no reason to believe in.

Richard's paradox and Borel's number are discussed in [Borel, 1950] on the pages given in the list of references, but the next paradox was considered so important by Borel that he devoted an entire book to it. In fact, this was Borel's last book [Borel, 1952] and it was published, as I said, when Borel was 81 years old. I think that when Borel wrote this work he must have been thinking about his legacy, since this was to be his final book-length mathematical statement. The Chinese, I believe, place special value on an artist's final work, considering that in some sense it contains or captures that artist's soul.[3] If so, [Borel, 1952] is Borel's "soul work."

Unfortunately I have not been able to obtain this crucial book. But based on a number of remarks by other people and based on what I do know about Borel's methods and concerns, I am fairly confident that I know what [Borel, 1952] contains. Here it is:[4]

2.5. Borel's "inaccessible numbers:" Most reals are un-nameable, with probability one

Borel's often-expressed credo is that a real number is really real only if it can be expressed, only if it can be uniquely defined, using a finite number of words.[5] It's only real if it can be named or specified as an individual mathematical object. And in order to do this we must necessarily employ some particular language, e.g., French. Whatever the choice of language,

---

[3] I certainly feel that way about Bach's *Die Kunst der Fuge* and about Bergman's *Fanny och Alexander*.

[4] *Note added in proof.* In fact, this is on page 21 of [Borel, 1952].

[5] See for example [Borel, 1960].

there will only be a countable infinity of possible texts, since these can be listed in size order, and among texts of the same size, in alphabetical order.

This has the devastating consequence that there are only a denumerable infinitely of such "accessible" reals, and therefore, as we saw in Sec. 2.2, the set of accessible reals has measure zero.

So, in Borel's view, most reals, with probability one, are mathematical fantasies, because there is no way to specify them uniquely. Most reals are inaccessible to us, and will never, ever, be picked out as individuals using **any** conceivable mathematical tool, because whatever these tools may be they could always be explained in French, and therefore can only "individualize" a countable infinity of reals, a set of reals of measure zero, an infinitesimal subset of the set of all possible reals.

Pick a real at random, and the probability is zero that it's accessible— the probability is zero that it **will ever be** accessible to us as an individual mathematical object.

3. History Repeats Itself: Computability Theory and Its Limitative Meta-Theorems

That was an exciting chapter in the history of ideas, wasn't it! But history moves on, and the collective attention of the human species shifts elsewhere, like a person who is examining a huge painting.

What completely transformed the situation is the **idea** of the computer, the computer as a mathematical concept, not a practical device, although the current ubiquity of computers doesn't hurt. It is, as usual, unfair to single out an individual, but in my opinion the crucial event was the 1936 paper by Turing *On computable numbers,* and here Turing is in fact referring to computable real numbers. You can find this paper at the beginning of the collection [Copeland, 2004], and at the end of this book there happens to be a much more understandable paper by Turing explaining just the key idea.[6]

History now repeats itself and recycles the ideas that were presented in Sec. 2. This time the texts will be written in artificial formal languages, they will be computer programs or proofs in a formal axiomatic math theory. They won't be texts that are written in a natural language like English or French.

---

[6]It's Turing's 1954 Penguin *Science News* paper on *Solvable and unsolvable problems,* which I copied out into a notebook by hand when I was a teenager.

And this time we won't get paradoxes, instead we'll get meta-theorems, we'll get limitative theorems, ones that show the limits of computation or the limitations of formal math theories. So in their current reincarnation, which we'll now present, the ideas that we saw in Sec. 2 definitely become much sharper and clearer.

Formal languages avoid the paradoxes by removing the ambiguities of natural languages. The paradoxes are eliminated, but there is a price. Paradoxical natural languages are evolving open systems. Artificial languages are static closed systems subject to limitative meta-theorems. You avoid the paradoxes, but you are left with a corpse!

The following tableau summarizes the transformation (paradigm shift):

- Natural languages $\longrightarrow$ Formal languages.

- Something is true $\longrightarrow$ Something is provable within a particular formal axiomatic math theory.[7]

- Naming a real number $\longrightarrow$ Computing a real number digit by digit.

- Number of words required to name something[8] $\longrightarrow$ Size in bits of the smallest program for computing something (program-size complexity).[9]

- List of all possible texts in French $\longrightarrow$ List of all possible programs, or
  List of all possible texts in French $\longrightarrow$ List of all possible proofs.[10]

- Paradoxes $\longrightarrow$ Limitative meta-theorems.

Now let's do Sec. 2 all over again. First we'll examine two different proofs that there are uncomputable reals: a diagonal argument proof, and a measure-theoretic proof. Then we'll show how the Richard paradox yields the unsolvability of the halting problem. Finally we'll discuss the halting probability Ω, which plays roughly the same role here that Borel's know-it-all real did in Sec. 2.

---

[7]This part of the paradigm shift is particularly important in the story of how Gödel converted the paradox of "this statement is false" into the proof of his famous 1931 incompleteness theorem, which is based on "this statement is unprovable." This changes something that's true if and only if it's false, into something that's true if and only if it's unprovable, thus transforming a paradox into a meta-theorem.

[8]See [Borel, 1960].

[9]See [Chaitin, 2005].

[10]The idea of systematically combining concepts in every possible way can be traced through Leibniz back to Ramon Llull (13th century), and is ridiculed by Swift in *Gulliver's Travels* (Part III, Chapter 5, on the Academy of Lagado).

3.1. Turing diagonalizes over all computable reals ⟶ uncomputable real

The set of all possible computer programs is countable, therefore the set of all computable reals is countable, and diagonalizing over the computable reals immediately yields an uncomputable real. *Q.E.D.*

Let's do it again more carefully.

Make a list of all possible computer programs. Order the programs by their size, and within those of the same size, order them alphabetically. The easiest thing to do is to include all the possible character strings that can be formed from the finite alphabet of the programming language, even though most of these will be syntactically invalid programs.

Here's how we define the uncomputable diagonal number $0 < r < 1$. Consider the kth program in our list. If it is syntactically invalid, or if the kth program never outputs a kth digit, or if the kth digit output by the kth program isn't a 3, pick 3 as the kth digit of r. Otherwise, if the kth digit output by the kth program is a 3, pick 4 as the kth digit of r.

This r cannot be computable, because its kth digit is different from the kth digit of the real number that is computed by the kth program, if there is one. Therefore there are uncomputable reals, real numbers that cannot be calculated digit by digit by any computer program.

3.2. Alternate proof: Reals are uncomputable with probability one

In a nutshell, the set of computer programs is countable, therefore the set of all computable reals is countable, and therefore, as in Sec. 2.2, of measure zero. *Q.E.D.*

More slowly, consider the kth computer program again. If it is syntactically invalid or fails to compute a real number, let's skip it. If it does compute a real, cover that real with an interval of length $\epsilon/2^k$. Then the total length of the covering is less than ϵ, which can be made arbitrarily small, and the computable reals are a null set.

In other words, the probability of a real's being computable is zero, and the probability that it's uncomputable is one.[11]

---

[11] Who should be credited for this measure-theoretic proof that there are uncomputable reals? I have no idea. It seems to have always been part of my mental baggage.

What if we allow arbitrary, highly nonconstructive means to specify particular reals, not just computer programs? The argument of Sec. 2.5 carries over immediately within our new framework in which we consider formal languages instead of natural languages. Most reals remain unnameable, with probability one.[12]

3.3. Turing's halting problem: No algorithm settles halting, no formal axiomatic math theory settles halting

Richard's paradox names an unnameable real. More precisely, it diagonalizes over all reals uniquely specified by French texts to produce a French text specifying an unspecifiable real. What becomes of this in our new context in which we name reals by computing them?

Let's go back to Turing's use of the diagonal argument in Sec. 3.1. In Sec. 3.1 we constructed an uncomputable real r. It must be uncomputable, by construction. Nevertheless, as was the case in the Richard paradox, it would seem that we gave a procedure for calculating Turing's diagonal real r digit by digit. How can this procedure fail? What could possibly go wrong?

The answer is this: The only noncomputable step has got to be determining if the kth computer program will **ever** output a kth digit. If we could do that, then we could certainly compute the uncomputable real r of Sec. 3.1.

In other words, Sec. 3.1 actually **proves** that there can be no algorithm for deciding if the kth computer program will ever output a kth digit.

And this is a special case of what's called Turing's halting problem. In this particular case, the question is whether or not the wait for a kth digit will ever terminate. In the general case, the question is whether or not a computer program will ever halt.

The algorithmic unsolvability of Turing's halting problem is an extremely fundamental meta-theorem. It's a much stronger result than Gödel's famous 1931 incompleteness theorem. Why? Because in Turing's original 1936 paper he immediately points out how to derive incompleteness from the halting problem.

A formal axiomatic math theory (FAMT) consists of a finite set of axioms and of a finite set of rules of inference for deducing the consequences of those axioms. Viewed from a great distance, all that counts is that there is an

---

[12]This theorem is featured in [Chaitin, 2005] at the end of the chapter entitled *The Labyrinth of the Continuum.*

algorithm for enumerating (or generating) all the possible theorems, all the possible consequences of the axioms, one by one, by systematically applying the rules of inference in every possible way. This is in fact what's called a breadth-first (rather than a depth-first) tree walk, the tree being the tree of all possible deductions.[13]

So, argued Turing in 1936, if there were a FAMT that always enabled you to decide whether or not a program eventually halts, there would in fact be an algorithm for doing so. You'd just run through all possible proofs until you find a proof that the program halts or you find a proof that it never halts.

So uncomputability is much more fundamental than incompleteness. Incompleteness is an immediate corollary of uncomputability. But uncomputability is **not** a corollary of incompleteness. The concept of incompleteness does not contain the concept of uncomputability.

Now let's get an even more disturbing limitative meta-theorem. We'll do that by considering the halting probability Ω [Chaitin, 2005], which is what corresponds to Borel's know-it-all real (Sec. 2.4) in the current context.[14]

3.4. Irreducible complexity, perfect randomness, maximal unknowability: The halting probability Ω

Where does the halting probability come from? Well, our motivation is the contrast between Sec. 3.1 and Sec. 3.2. Sec. 3.1 is to Sec. 3.2 as the halting problem is to the halting probability! In other words, the fact that we found an easier way to show the existence of uncomputable reals using a probabilistic argument, suggests looking at the probability that a program chosen at random will ever halt instead of considering individual programs as in Turing's 1936 paper.

Formally, the halting probability Ω is defined as follows:

$$0 \; < \; \Omega \; \equiv \sum_{\text{program } p \text{ halts}} 2^{-(\text{the size in bits of } p)} \; < \; 1.$$

To avoid having this sum diverge to infinity instead of converging to a number between zero and one, it is important that the programs p should be self-

---

[13]This is another way to achieve the effect of running through all possible texts.

[14][Tasić, 2001] was the first person to make the connection between Borel's real and Ω. I became aware of Borel's real through Tasić.

delimiting (no extension of a valid program is a valid program; see [Chaitin, 2005]).

What's interesting about Ω is that it behaves like a compressed version of Borel's know-it-all real. Knowing the first n bits of Borel's real enables us to answer the first n yes/no questions in French. Knowing the first n bits of Ω enables us to answer the halting problem for all programs p up to n bits in size. I.e., n bits of Ω tells us whether or not each p up to n bits in size ever halts. (Can you see how?) That's a lot of information!

In fact, Ω compactly encodes so much information that you essentially need an n-bit FAMT in order to be able to determine n bits of Ω! In other words, Ω is **irreducible mathematical information**, it's a place where reasoning is completely impotent. The bits of Ω are mathematical facts that can be proved, but essentially only by adding them one by one as new axioms! I'm talking about how difficult it is to prove theorems such as

> "the 5th bit of Ω is a 0"

and

> "the 9th bit of Ω is a 1"

or whatever the case may be.

To prove that Ω is computationally and therefore logically irreducible, requires a theory of program-size complexity that I call algorithmic information theory (AIT) [Chaitin, 2005]. The key idea in AIT is to measure the complexity of something via the size in bits of the smallest program for calculating it. This is a more refined version of Borel's idea [Borel, 1960] of defining the complexity of a real number to be the number of words required to name it.

And the key fact that is proved in AIT about Ω is that

$$H(\Omega_n) \geq n - c.$$

I.e.,

> (the string Ω_n consisting of the first n bits of Ω)

has program-size complexity or "algorithmic entropy H" greater than or equal to $n - c$. Here c is a constant, and I'm talking about the size in bits of self-delimiting programs.

In other words, any self-delimiting program for computing the first n bits of Ω will have to be at least $n - c$ bits long.

The irreducible sequence of bits of Ω is a place where mathematical truth has absolutely no pattern or structure that we will ever be able to detect. It's a place where mathematical truth has maximum possible entropy—a place where, in a sense, God plays dice.[15]

Why should we believe in real numbers, if most of them are uncomputable? Why should we believe in real numbers, if most of them, it turns out,[16] are maximally unknowable like Ω?[17]

4. Digital Philosophy and Digital Physics

So much for mathematics! Now let's turn to physics.

Discreteness entered modern science through chemistry, when it was discovered that matter is built up out of atoms and molecules. Recall that the first experimental evidence for this was Gay-Lussac's discovery of the simple integer ratios between the volumes of gaseous substances that are combined in chemical reactions. This was the first evidence, two centuries ago, that discreteness plays an important role in the physical world.

At first it might seem that quantum mechanics (QM), which began with Einstein's photon as the explanation for the photoelectric effect in 1905, goes further in the direction of discreteness. But the wave-particle duality discovered by de Broglie in 1925 is at the heart of QM, which means that this theory is profoundly ambiguous regarding the question of discreteness vs. continuity. QM can have its cake and eat it too, because discreteness is modeled via standing waves (eigenfunctions) in a continuous medium.

The latest strong hints in the direction of discreteness come from quantum gravity [Smolin, 2000], in particular from the Bekenstein bound and the so-called "holographic principle." According to these ideas the amount of information in any physical system is bounded, i.e., is a finite number of 0/1 bits.

---

[15]On the other hand, if Gödel is correct in thinking that mathematical intuition can at times directly perceive the Platonic world of mathematical ideas, then the bits of Ω may in fact be accessible.

[16]See the chapter entitled *The Labyrinth of the Continuum* in [Chaitin, 2005].

[17]In spite of the fact that most individual real numbers will forever escape us, the notion of an arbitrary real has beautiful mathematical properties and is a concept that helps us to organize and understand the real world. Individual concepts in a theory **do not** need to have concrete meaning on their own; it is enough if the theory **as a whole** can be compared with the results of experiments.

But it is not just fundamental physics that is pushing us in this direction. Other hints come from our pervasive digital technology, from molecular biology where DNA is the digital software for life, and from *a priori* philosophical prejudices going back to the ancient Greeks.

According to Pythagoras everything is number, and God is a mathematician. This point of view has worked pretty well throughout the development of modern science. However now a neo-Pythagorian doctrine is emerging, according to which everything is 0/1 bits, and the world is built entirely out of digital information. In other words, now everything is software, God is a computer programmer, not a mathematician, and the world is a giant information-processing system, a giant computer [Fredkin, 2004, Wolfram, 2002, Chaitin, 2005].

Indeed, the most important thing in understanding a complex system is to understand how it processes information. This viewpoint regards physical systems as information processors, as performing computations. This approach also sheds new light on microscopic quantum systems, as is demonstrated in the highly developed field of quantum information and quantum computation. An extreme version of this doctrine would attempt to build the world entirely out of discrete digital information, out of 0 and 1 bits.[18]

Whether or not this ambitious new research program can eventually succeed, it will be interesting to see how far it gets. The problem of the infinite divisibility of space and time has been with us for more than two millennia, since Zeno of Elea and his famous paradoxes, and it is also discussed by Maimonides in his *Guide for the Perplexed* (12th century).

Modern versions of this ancient problem are, for example, the infinite amount of energy contained in the electric field surrounding a point electron according to Maxwell's theory of electromagnetism, and the breakdown of space-time because of the formation of black holes due to extreme quantum fluctuations (arbitrarily high energy virtual pairs) in the vacuum quantum field.

I do not expect that the tension between the continuous and the discrete will be resolved any time soon. Nevertheless, one must try. And, as we have seen in our two case studies, before being swept away, each generation contributes something to the ongoing discussion.

---

[18]This idea, like so many others, can be traced back to Leibniz. He thought it was important enough to have it cast in the form of a medallion.

References

Borel, E. [1950] *Leçons sur la Théorie des Fonctions* (Gabay, Paris) pp. 161, 275.

Borel, E. [1952] *Les Nombres Inaccessibles* (Gauthier-Villars, Paris) p. 21.

Borel, E. [1960] *Space and Time* (Dover, Mineola) pp. 212–214.

Chaitin, G. [2005] *Meta Math!* (Pantheon, New York).

Copeland, B. J. [2004] *The Essential Turing* (Clarendon Press, Oxford).

Courant, R. & Robbins, H. [1947] *What is Mathematics?* (Oxford University Press, New York) Sec. II.4.2, pp. 79–83.

Fredkin, E. [2004] http://digitalphilosophy.org.

James, I. [2002] *Remarkable Mathematicians* (Cambridge University Press, Cambridge) pp. 283–292.

Smolin, L. [2000] *Three Roads to Quantum Gravity* (Weidenfeld & Nicolson, London) Chap. 8, pp. 95–105, Chap. 12, pp. 169–178.

Tasić, V. [2001] *Mathematics and the Roots of Postmodern Thought* (Oxford University Press, New York) pp. 52, 81–82.

Wolfram, S. [2002] *A New Kind of Science* (Wolfram Media, Champaign).

Epistemology as information theory: From Leibniz to Ω

*In 1686 in his **Discours de métaphysique,** Leibniz points out that if an arbitrarily complex theory is permitted then the notion of "theory" becomes vacuous because there is always a theory. This idea is developed in the modern theory of algorithmic information, which deals with the size of computer programs and provides a new view of Gödel's work on incompleteness and Turing's work on uncomputability. Of particular interest is the halting probability Ω, whose bits are irreducible, i.e., maximally unknowable mathematical facts. More generally, these ideas constitute a kind of "digital philosophy" related to recent attempts of Edward Fredkin, Stephen Wolfram and others to view the world as a giant computer. There are also connections with recent "digital physics" speculations that the universe might actually be discrete, not continuous. This **système du monde** is presented as a coherent whole in my book **Meta Math!,** which will be published this fall. [Alan Turing Lecture on Computing and Philosophy, E-CAP'05, European Computing and Philosophy Conference, Mälardalen University, Västerås, Sweden, June 2005.]*

Introduction

I am happy to be here with you enjoying the delicate Scandinavian summer; if we were a little farther north there wouldn't be any darkness at all. And I am especially delighted to be here delivering the Alan Turing Lecture. Turing's famous 1936 paper is an intellectual milestone that seems larger and more important with every passing year.[1]

People are not merely content to enjoy the beautiful summers in the far north, they also want and need **to understand**, and so they create myths. In this part of the world those myths involve Thor and Odin and the other Norse gods. In this talk, I'm going to present another myth, what the French call a *système du monde,* a system of the world, a speculative metaphysics based on information and the computer.[2]

The previous century had logical positivism and all that emphasis on the philosophy of language, and completely shunned speculative metaphysics, but a number of us think that it is time to start again. There is an emerging digital philosophy and digital physics, a new metaphysics associated with names like Edward Fredkin and Stephen Wolfram and a handful of like-minded individuals, among whom I include myself. As far as I know the terms "digital philosophy" and "digital physics" were actually invented by Fredkin, and he has a large website with his papers and a draft of a book about this. Stephen Wolfram attracted a great deal of attention to the movement and stirred up quite a bit of controversy with his very large and idiosyncratic book on *A New Kind of Science.*

And I have my own book on the subject, in which I've attempted to wrap everything I know and care about into a single package. It's a small book, and amazingly enough it's going to be published by a major New York publisher a few months from now. This talk will be an overview of my book, which presents my own personal version of "digital philosophy," since each of us who works in this area has a different vision of this tentative, emerging world view. My book is called *Meta Math!,* which may not seem like a serious title, but it's actually a book intended for my professional colleagues as well as for the general public, the high-level, intellectual, thinking public.

"Digital philosophy" is actually a neo-Pythagorean vision of the world, it's just a new version of that. According to Pythagoras, all is number — and by number he means the positive integers, 1, 2, 3, ... — and God is a mathematician. "Digital philosophy" updates this as follows: Now everything is

---

[1]For Turing's original paper, with commentary, see Copeland's *The Essential Turing.*

[2]One reader's reaction (GDC): "Grand unified theories may be like myths, but surely there is a difference between scientific theory and any other narrative?" I would argue that a scientific narrative is more successful than the Norse myths because it explains what it explains more precisely and without having to postulate new gods all the time, i.e., it's a better "compression" (which will be my main point in this lecture; that's how you measure how successful a theory is).

made out of 0/1 bits, everything is digital software, and God is a computer programmer, not a mathematician! It will be interesting to see how well this vision of the world succeeds, and just how much of our experience and theorizing can be included or shoe-horned within this new viewpoint.[3]

Let me return now to Turing's famous 1936 paper. This paper is usually remembered for inventing the programmable digital computer via a mathematical model, the Turing machine, and for discovering the extremely fundamental halting problem. Actually Turing's paper is called "On computable numbers, with an application to the *Entscheidungsproblem*," and by computable numbers Turing means "real" numbers, numbers like e or $\pi = 3.1415926\ldots$ that are measured with infinite precision, and that can be computed with arbitrarily high precision, digit by digit without ever stopping, on a computer.

Why do I think that Turing's paper "On computable numbers" is so important? Well, in my opinion it's a paper on epistemology, because we only understand something if we can program it, as I will explain in more detail later. And it's a paper on physics, because what we can actually compute depends on the laws of physics in our particular universe and distinguishes it from other possible universes. And it's a paper on ontology, because it shows that some real numbers are **uncomputable**, which I shall argue calls into question their very existence, their mathematical and physical existence.[4]

To show how strange uncomputable real numbers can be, let me give a particularly illuminating example of one, which actually preceded Turing's

---

[3]Of course, a system of the world can only work by omitting everything that doesn't fit within its vision. The question is how much will fail to fit, and conversely, how many things will this vision be able to help us to understand. Remember, if one is wearing rose colored glasses, everything seems pink. And as Picasso said, theories are lies that help us to see the truth. No theory is perfect, and it will be interesting to see how far this digital vision of the world will be able to go.

[4]You might exclaim (GDC), "You can't be saying that before Turing and the computer no one understood anything; that can't be right!" My response to this is that before Turing (and my theory) people could understand things, but **they couldn't measure how well** they understood them. Now you can measure that, in terms of the degree of compression that is achieved. I will explain this later at the beginning of the section on computer epistemology. Furthermore, programming something forces you to understand it better, it forces you to really understand it, since you are explaining it **to a machine**. That's sort of what happens when a student or a small child asks you what at first you take to be a stupid question, and then you realize that this question has in fact done you the favor of forcing you to formulate your ideas more clearly and perhaps even question some of your tacit assumptions.

1936 paper. It's a very strange number that was invented in a 1927 paper by the French mathematician Emile Borel. Borel's number is sort of an anticipation, a partial anticipation, of Turing's 1936 paper, but that's only something that one can realize in retrospect. Borel presages Turing, which does not in any way lessen Turing's important contribution that so dramatically and sharply clarified all these vague ideas.[5]

Borel was interested in "constructive" mathematics, in what you can actually compute we would say nowadays. And he came up with an extremely strange non-constructive real number. You list all possible yes/no questions in French in an immense, an infinite list of all possibilities. This will be what mathematicians call a denumerable or a countable infinity of questions, because it can be put into a one-to-one correspondence with the list of positive integers 1, 2, 3, ... In other words, there will be a first question, a second question, a third question, and in general an Nth question.

You can imagine all the possible questions to be ordered by size, and within questions of the same size, in alphabetical order. More precisely, you consider all possible strings, all possible finite sequences of symbols in the French alphabet, including the blank so that you get words, and the period so that you have sentences. And you imagine filtering out all the garbage and being left only with grammatical yes/no questions in French. Later I will tell you in more detail how to actually do this. Anyway, for now **imagine** doing this, and so there will be a first question, a second question, an Nth question.

And the Nth digit or the Nth bit after the decimal point of Borel's number answers the Nth question: It will be a 0 if the answer is no, and it'll be a 1 if the answer is yes. So the binary expansion of Borel's number contains the answer to every possible yes/no question! It's like having an oracle, a Delphic oracle that will answer every yes/no question!

How is this possible?! Well, according to Borel, it isn't really possible, this can't be, it's totally unbelievable. This number is only a mathematical fantasy, it's not for real, it cannot claim a legitimate place in our ontology. Later I'll show you a modern version of Borel's number, my halting probability Ω. And I'll tell you why some contemporary physicists, real physicists, not mavericks, are moving in the direction of digital physics.

[Actually, to make Borel's number as real as possible, you have to avoid the

---

[5]I learnt of Borel's number by reading Tasic's *Mathematics and the Roots of Postmodern Thought*, which also deals with many of the issues discussed here.

problem of filtering out all the yes/no questions. And you have to use decimal digits, you can't use binary digits. You number all the possible finite strings of French symbols including blanks and periods, which is quite easy to do using a computer. Then the Nth digit of Borel's number is 0 if the Nth string of characters in French is ungrammatical and not proper French, it's 1 if it's grammatical, but not a yes/no question, it's 2 if it's a yes/no question that cannot be answered (e.g., "Is the answer to this question 'no'?"), it's 3 if the answer is no, and it's 4 if the answer is yes.]

Geometrically a real number is the most straightforward thing in the world, it's just a point on a line. That's quite natural and intuitive. But *arithmetically,* that's another matter. The situation is quite different. From an arithmetical point of view reals are extremely problematical, they are fraught with difficulties!

Before discussing my Ω number, I want to return to the fundamental question of what does it mean to understand. How do we explain or comprehend something? What is a theory? How can we tell whether or not it's a successful theory? How can we measure how successful it is? Well, using the ideas of information and computation, that's not difficult to do, and the central idea can even be traced back to Leibniz's 1686 *Discours de métaphysique.*

Computer Epistemology: What is a mathematical or scientific theory? How can we judge whether it works or not?

In Sections V and VI of his *Discourse on Metaphysics*, Leibniz asserts that God simultaneously maximizes the variety, diversity and richness of the world, and minimizes the conceptual complexity of the set of ideas that determine the world. And he points out that for any finite set of points there is always a mathematical equation that goes through them, in other words, a law that determines their positions. But if the points are chosen at random, that equation will be extremely complex.

This theme is taken up again in 1932 by Hermann Weyl in his book *The Open World* consisting of three lectures he gave at Yale University on the metaphysics of modern science. Weyl formulates Leibniz's crucial idea in the following extremely dramatic fashion: If one permits arbitrarily complex laws, then the concept of law becomes vacuous, because there is always a law! Then Weyl asks, how can we make more precise the distinction between mathematical simplicity and mathematical complexity? It seems to be very hard to do that. How can we measure this important parameter, without which it is impossible to distinguish between a successful theory and one that is completely unsuccessful?

This problem is taken up and I think satisfactorily resolved in the new mathematical theory I call *algorithmic information theory*. The epistemological model that is central to this theory is that a scientific or mathematical theory is a computer program for calculating the facts, and the smaller the program, the better. The complexity of your theory, of your law, is measured in bits of software:

program (bit string) $\longrightarrow$ **Computer** $\longrightarrow$ output (bit string)
theory $\longrightarrow$ **Computer** $\longrightarrow$ mathematical or scientific facts
Understanding is compression!

Now Leibniz's crucial observation can be formulated much more precisely. For any finite set of scientific or mathematical facts, there is always a theory that is exactly as complicated, exactly the same size in bits, as the facts themselves. (It just directly outputs them "as is," without doing any computation.) But that doesn't count, that doesn't enable us to distinguish between what can be comprehended and what cannot, because there is always a theory that is as complicated as what it explains. A theory, an explanation, is only successful to the extent to which it compresses the number of bits in the facts into a much smaller number of bits of theory. Understanding is compression, comprehension is compression! That's how we can tell the difference between real theories and *ad hoc* theories.[6]

What can we do with this idea that an explanation has to be simpler than what it explains? Well, the most important application of these ideas that I have been able to find is in metamathematics, it's in discussing what mathematics can or cannot achieve. You simultaneously get an information-theoretic, computational perspective on Gödel's famous 1931 incompleteness theorem, and on Turing's famous 1936 halting problem. How?[7]

---

[6]By the way, Leibniz also mentions complexity in Section 7 of his *Principles of Nature and Grace*, where he asks the amazing question, "Why is there something rather than nothing? For nothing is simpler and easier than something."

[7]For an insightful treatment of Gödel as a philosopher, see Rebecca Goldstein's *Incompleteness*.

Here's how! These are my two favorite information-theoretic incompleteness results:

- You need an N-bit theory in order to be able to prove that a specific N-bit program is "elegant."

- You need an N-bit theory in order to be able to determine N bits of the numerical value, of the base-two binary expansion, of the halting probability Ω.

Let me explain.

What is an elegant program? It's a program with the property that no program written in the same programming language that produces the same output is smaller than it is. In other words, an elegant program is the most concise, the simplest, the best theory for its output. And there are infinitely many such programs, they can be arbitrarily big, because for any computational task there has to be at least one elegant program. (There may be several if there are ties, if there are several programs for the same output that have exactly the minimum possible number of bits.)

And what is the halting probability Ω? Well, it's defined to be the probability that a computer program generated at random, by choosing each of its bits using an independent toss of a fair coin, will eventually halt. Turing is interested in whether or not individual programs halt. I am interested in trying to prove what are the bits, what is the numerical value, of the halting probability Ω. By the way, the value of Ω depends on your particular choice of programming language, which I don't have time to discuss now. Ω is also equal to the result of summing $1/2$ raised to powers which are the size in bits of every program that halts. In other words, each K-bit program that halts contributes $1/2^K$ to Ω.

And what precisely do I mean by an N-bit mathematical theory? Well, I'm thinking of formal axiomatic theories, which are formulated using symbolic logic, not in any natural, human language. In such theories there are always a finite number of axioms and there are explicit rules for mechanically deducing consequences of the axioms, which are called theorems. An N-bit theory is one for which there is an N-bit program for systematically running through the tree of all possible proofs deducing all the consequences of the axioms, which are all the theorems in your formal theory. This is slow work, but in principle it can be done mechanically, that's what counts. David Hilbert believed that there had to be a single formal axiomatic theory for all of mathematics; that's just another way of stating that math is static and perfect and provides absolute truth.

Not only is this impossible, not only is Hilbert's dream impossible to achieve, but there are in fact an infinity of irreducible mathematical truths, mathematical truths for which essentially the only way to prove them is to add them as new axioms. My first example of such truths was determining elegant programs, and an even better example is provided by the bits of Ω. The bits of Ω are mathematical facts that are true for no reason (no reason simpler than themselves), and thus violate Leibniz's principle of sufficient reason, which states that if anything is true it has to be true for a reason.

In math the reason that something is true is called its proof. Why are the bits of Ω true for no reason, why can't you prove what their values are? Because, as Leibniz himself points out in Sections 33 to 35 of *The Monadology,* the essence of the notion of proof is that you prove a complicated assertion by analyzing it, by breaking it down until you reduce its truth to the truth of assertions that are so simple that they no longer require any proof (self-evident axioms). But if you cannot deduce the truth of something from any principle simpler than itself, then proofs become useless, because **anything** can be proven from principles that are equally complicated, e.g., by directly adding it as a new axiom without any proof. And this is exactly what happens with the bits of Ω.

In other words, the normal, Hilbertian view of math is that all of mathematical truth, an infinite number of truths, can be compressed into a finite number of axioms. But there are an infinity of mathematical truths that cannot be compressed at all, not one bit!

This is an amazing result, and I think that it has to have profound philosophical and practical implications. Let me try to tell you why.

On the one hand, it suggests that pure math is more like biology than it is like physics. In biology we deal with very complicated organisms and mechanisms, but in physics it is normally assumed that there has to be a theory of everything, a simple set of equations that would fit on a T-shirt and in principle explains the world, at least the physical world. But we have seen that the world of mathematical ideas has infinite complexity, it cannot be explained with any theory having a finite number of bits, which from a sufficiently abstract point of view seems much more like biology, the domain of the complex, than like physics, where simple equations reign supreme.

On the other hand, this amazing result suggests that even though math and physics are different, they may not be as different as most people think! I mean this in the following sense: In math you organize your computational experience, your lab is the computer, and in physics you organize physical experience and have real labs. But in both cases an explanation has to be simpler than what it explains, and in both cases there are sets of facts that cannot be explained, that are irreducible. Why? Well, in quantum physics it is assumed that there are phenomena that when measured are equally likely to give either of two answers (e.g., spin up, spin down) and that are inherently unpredictable and irreducible. And in pure math we have a similar example, which is provided by the individual bits in the binary expansion of the numerical value of the halting probability Ω.

This suggests to me a quasi-empirical view of math, in which one is more willing to add new axioms that are not at all self-evident but that are justified pragmatically, i.e., by their fruitful consequences, just like a physicist would. I have taken the term quasi-empirical from Lakatos. The collection of essays *New Directions in the Philosophy of Mathematics* edited by Tymoczko in my opinion pushes strongly in the direction of a quasi-empirical view of math, and it contains an essay by Lakatos proposing the term "quasi-empirical," as well as essays of my own and by a number of other people. Many of them may disagree with me, and I'm sure do, but I repeat, in my opinion all of these essays justify a quasi-empirical view of math, what I mean by quasi-empirical, which is somewhat different from what Lakatos originally meant, but is in quite the same spirit, I think.

In a two-volume work full of important mathematical examples, Borwein, Bailey and Girgensohn have argued that experimental mathematics is an extremely valuable research paradigm that should be openly acknowledged and indeed vigorously embraced. They do not go so far as to suggest that one should add new axioms whenever they are helpful, without bothering with proofs, but they are certainly going in that direction and nod approvingly at my attempts to provide some theoretical justification for their entire enterprise by arguing that math and physics are not that different.

In fact, since I began to espouse these heretical views in the early 1970's, largely to deaf ears, there have actually been several examples of such new pragmatically justified, non-self-evident axioms:

- the **P** ≠ **NP** hypothesis regarding the time complexity of computations,

- the axiom of projective determinacy in set theory, and

- increasing reliance on diverse unproved versions of the Riemann hypothesis regarding the distribution of the primes.

So people don't need to have theoretical justification; they just do whatever is needed to get the job done...

The only problem with this computational and information-theoretic epistemology that I've just outlined to you is that it's based on the computer, and there are uncomputable reals. So what do we do with contemporary physics which is full of partial differential equations and field theories, all of which are formulated in terms of real numbers, **most of which are in fact uncomputable**, as I'll now show. Well, it would be good to get rid of all that and convert to a *digital physics*. Might this in fact be possible?! I'll discuss that too.

Computer Ontology: How real are real numbers? What is the world made of?

How did Turing prove that there are uncomputable reals in 1936? He did it like this. Recall that the possible texts in French are a countable or denumerable infinity and can be placed in an infinite list in which there is a first one, a second one, etc. Now let's do the same thing with all the possible computer programs (first you have to choose your programming language). So there is a first program, a second program, etc. Every computable real can be calculated digit by digit by some program in this list of all possible programs. Write the numerical value of that real next to the programs that calculate it, and cross off the list all the programs that do not calculate an individual computable real. We have converted a list of programs into a list of computable reals, and no computable real is missing.

Next discard the integer parts of all these computable reals, and just keep the decimal expansions. Then put together a new real number by changing every digit on the diagonal of this list (this is called Cantor's diagonal method; it comes from set theory). So your new number's first digit differs from the first digit of the first computable real, its second digit differs from the second digit of the second computable real, its third digit differs from the third digit of the third computable real, and so forth and so on. So it can't be in the list of all computable reals and it has to be uncomputable.

And that's Turing's uncomputable real number![8]

Actually, there is a much easier way to see that there are uncomputable reals by using ideas that go back to Emile Borel (again!). Technically, the argument that I'll now present uses what mathematicians call *measure theory,* which deals with probabilities. So let's just look at all the real numbers between 0 and 1. These correspond to points on a line, a line exactly one unit in length, whose leftmost point is the number 0 and whose rightmost point is the number 1. The total length of this line segment is of course exactly one unit. But I will now show you that all the computable reals in this line segment can be covered using intervals whose total length can be made as small as desired. In technical terms, the computable reals in the interval from 0 to 1 are a set of measure zero, they have zero probability.

How do you cover all the computable reals? Well, remember that list of all the computable reals that we just diagonalized over to get Turing's uncomputable real? This time let's cover the first computable real with an interval of size $\epsilon/2$, let's cover the second computable real with an interval of size $\epsilon/4$, and in general we'll cover the Nth computable real with an interval of size $\epsilon/2^N$. The total length of all these intervals (which can conceivably overlap or fall partially outside the unit interval from 0 to 1), is exactly equal to ϵ, which can be made as small as we wish! In other words, there are arbitrarily small coverings, and the computable reals are therefore a set of measure zero, they have zero probability, they constitute an infinitesimal fraction of all the reals between 0 and 1. So if you pick a real at random between 0 and 1, with a uniform distribution of probability, it is infinitely unlikely, though possible, that you will get a computable real!

What disturbing news! Uncomputable reals are not the exception, they are the majority! How strange!

In fact, the situation is even worse than that. As Emile Borel points out on page 21 of his final book, *Les nombres inaccessibles* (1952), without making any reference to Turing, most individual reals are not even uniquely specifiable, they cannot even be named or pointed out, no matter how non-constructively, because of the limitations of human languages, which permit only a countable infinity of possible texts. The individually accessible or nameable reals are also a set of measure zero. Most reals are un-nameable, with probability one! I rediscovered this result of Borel's on my own in a

---

[8] *Technical Note:* Because of **synonyms** like $.345999\ldots = .346000\ldots$ you should avoid having any 0 or 9 digits in Turing's number.

slightly different context, in which things can be done a little more rigorously, which is when one is dealing with a formal axiomatic theory or an **artificial** formal language instead of a natural human language. That's how I present this idea in *Meta Math!*.

So if most individual reals will forever escape us, why should we believe in them?! Well, you will say, because they have a pretty structure and are a nice theory, a nice game to play, with which I certainly agree, and also because they have important practical applications, they are needed in physics. Well, perhaps not! Perhaps physics can give up infinite precision reals! How? Why should physicists want to do that?

Because it turns out that there are actually many reasons for being skeptical about the reals, in classical physics, in quantum physics, and particularly in more speculative contemporary efforts to cobble together a theory of black holes and quantum gravity.

First of all, as my late colleague the physicist Rolf Landauer used to remind me, no physical measurement has ever achieved more than a small number of digits of precision, not more than, say, 15 or 20 digits at most, and such high-precision experiments are rare masterpieces of the experimenter's art and not at all easy to achieve.

This is only a practical limitation in classical physics. But in quantum physics it is a consequence of the Heisenberg uncertainty principle and wave-particle duality (de Broglie). According to quantum theory, the more accurately you try to measure something, the smaller the length scales you are trying to explore, the higher the energy you need (the formula describing this involves Planck's constant). That's why it is getting more and more expensive to build particle accelerators like the one at CERN and at Fermilab, and governments are running out of money to fund high-energy physics, leading to a paucity of new experimental data to inspire theoreticians.

Hopefully new physics will eventually emerge from astronomical observations of bizarre new astrophysical phenomena, since we have run out of money here on earth! In fact, currently some of the most interesting physical speculations involve the thermodynamics of black holes, massive concentrations of matter that seem to be lurking at the hearts of most galaxies. Work by Stephen Hawking and Jacob Bekenstein on the thermodynamics of black holes suggests that any physical system can contain only a finite amount of information, a finite number of bits whose possible maximum is determined by what is called the Bekenstein bound. Strangely enough, this bound on the number of bits grows as the surface area of the physical system, not as its volume, leading to the so-called "holographic" principle asserting that in some sense space is actually two-dimensional even though it appears to have three dimensions!

So perhaps continuity is an illusion, perhaps everything is really discrete. There is another argument against the continuum if you go down to what is called the Planck scale. At distances that extremely short our current physics breaks down because spontaneous fluctuations in the quantum vacuum should produce mini-black holes that completely tear spacetime apart. And that is not at all what we see happening around us. So perhaps distances that small **do not exist**.

Inspired by ideas like this, in addition to *a priori* metaphysical biases in favor of discreteness, a number of contemporary physicists have proposed building the world out of discrete information, out of bits. Some names that come to mind in this connection are John Wheeler, Anton Zeilinger, Gerard 't Hooft, Lee Smolin, Seth Lloyd, Paola Zizzi, Jarmo Mäkelä and Ted Jacobson, who are real physicists. There is also more speculative work by a small cadre of cellular automata and computer enthusiasts including Edward Fredkin and Stephen Wolfram, whom I already mentioned, as well as Tommaso Toffoli, Norman Margolus, and others.

And there is also an increasing body of highly successful work on quantum computation and quantum information that is not at all speculative, it is just a fundamental reworking of standard 1920's quantum mechanics. Whether or not quantum computers ever become practical, the workers in this highly popular field have clearly established that it is illuminating to study sub-atomic quantum systems in terms of how they process qubits of quantum information and how they perform computation with these qubits. These notions have shed completely new light on the behavior of quantum mechanical systems.

Furthermore, when dealing with complex systems such as those that occur in biology, thinking about information processing is also crucial. As I believe Seth Lloyd said, the most important thing in understanding a complex system is to determine how it represents information and how it processes that information, i.e., what kinds of computations are performed.

And how about the entire universe, can it be considered to be a computer? Yes, it certainly can, it is constantly computing its future state from its current state, it's constantly computing its own time-evolution! And as I believe Tom Toffoli pointed out, actual computers like your PC just hitch a ride on this universal computation!

So perhaps we are not doing violence to Nature by attempting to force her into a digital, computational framework. Perhaps she has been flirting with us, giving us hints all along, that she is really discrete, not continuous, hints that we choose not to hear, because we are so much in love and don't want her to change!

For more on this kind of new physics, see the books by Smolin and von Baeyer in the bibliography. Several more technical papers on this subject are also included there.

Conclusion

Let me now wrap this up and try to give you a present to take home, more precisely, a piece of homework. In extremely abstract terms, I would say that the problem is, as was emphasized by Ernst Mayr in his book *This is Biology,* that the current philosophy of science deals more with physics and mathematics than it does with biology. But let me try to put this in more concrete terms and connect it with the spine, with the central thread, of the ideas in this talk.

To put it bluntly, a closed, static, eternal fixed view of math can no longer be sustained. As I try to illustrate with examples in my *Meta Math!* book, math actually advances by inventing new concepts, by completely changing the viewpoint. Here I emphasized new axioms, increased complexity, more information, but what really counts are new ideas, new concepts, new viewpoints. And that leads me to the crucial question, crucial for a proper open, dynamic, time-dependent view of mathematics,

"Where do new mathematical ideas come from?"

I repeat, math does not advance by mindlessly and mechanically grinding away deducing all the consequences of a fixed set of concepts and axioms, not at all! It advances with new concepts, new definitions, new perspectives, through revolutionary change, paradigm shifts, not just by hard work.

In fact, I believe that this is actually the central question in biology as well as in mathematics, it's the mystery of creation, of creativity:

"Where do new mathematical and biological ideas come from?"
"How do they emerge?"

Normally one equates a new biological idea with a new species, but in fact every time a child is born, that's actually a new idea incarnating; it's reinventing the notion of "human being," which changes constantly.

I have no idea how to answer this extremely important question; I wish I could. Maybe **you** will be able to do it. Just try! You might have to keep it cooking on a back burner while concentrating on other things, but don't give up! All it takes is a new idea! Somebody has to come up with it. Why not you?[9]

Appendix: Leibniz and the Law

I am indebted to Professor Ugo Pagallo for explaining to me that Leibniz, whose ideas and their elaboration were the subject of my talk, is regarded as just as important in the field of law as he is in the fields of mathematics and philosophy.

The theme of my lecture was that if a law is arbitrarily complicated, then it is not a law; this idea was traced via Hermann Weyl back to Leibniz. In mathematics it leads to my Ω number and the surprising discovery of completely lawless regions of mathematics, areas in which there is absolutely no structure or pattern or way to understand what is happening.

The principle that an arbitrarily complicated law is not a law can also be

---

[9]I'm not denying the importance of Darwin's theory of evolution. But I want much more than that, I want a profound, extremely general mathematical theory that captures the essence of what life is and why it evolves. I want a theory that gets to the heart of the matter. And I suspect that any such theory will necessarily have to shed new light on mathematical creativity as well. Conversely, a deep theory of mathematical creation might also cover biological creativity.

A reaction from Gordana Dodig-Crnkovic: "Regarding Darwin and Neo-Darwinism I agree with you — it is a very good idea to go beyond. In my view there is nothing more beautiful and convincing than a good mathematical theory. And I do believe that it must be possible to express those thoughts in a much more general way... I believe that it is a very crucial thing to try to formulate life in terms of computation. Not to say life is nothing more than a computation. But just to explore how far one can go with that idea. Computation seems to me a very powerful tool to illuminate many things about the material world and the material ground for mental phenomena (including creativity)... Or would you suggest that creativity is given by God's will? That it is the very basic axiom? Isn't it possible to relate to pure chance? Chance and selection? Wouldn't it be a good idea to assume two principles: law and chance, where both are needed to reconstruct the universe in computational terms? (like chaos and cosmos?)"

interpreted with reference to the legal system. It is not a coincidence that the words "law" and "proof" and "evidence" are used in jurisprudence as well as in science and mathematics. In other words, the rule of law is equivalent to the rule of reason, but if a law is sufficiently complicated, then it can in fact be completely arbitrary and incomprehensible.

Acknowledgements

I wish to thank Gordana Dodig-Crnkovic for organizing E-CAP'05 and for inviting me to present the Turing lecture at E-CAP'05; also for stimulating discussions reflected in those footnotes that are marked with GDC. The remarks on biology are the product of a week spent in residence at Rockefeller University in Manhattan, June 2005; I thank Albert Libchaber for inviting me to give a series of lectures there to physicists and biologists. The appendix is the result of lectures to philosophy of law students April 2005 at the Universities of Padua, Bologna and Turin; I thank Ugo Pagallo for arranging this. Thanks too to Paola Zizzi for help with the physics references.

References

- Edward Fredkin, http://digitalphilosophy.org.
- Stephen Wolfram, *A New Kind of Science*, Wolfram Media, 2002.
- Gregory Chaitin, *Meta Math!*, Pantheon, 2005.
- G. W. Leibniz, *Discourse on Metaphysics, Principles of Nature and Grace, The Monadology*, 1686, 1714, 1714.
- Hermann Weyl, *The Open World*, Yale University Press, 1932.
- Thomas Tymoczko, *New Directions in the Philosophy of Mathematics*, Princeton University Press, 1998.
- Jonathan Borwein, David Bailey, Roland Girgensohn, *Mathematics by Experiment, Experimentation in Mathematics*, A. K. Peters, 2003, 2004.
- Rebecca Goldstein, *Incompleteness*, Norton, 2005.
- B. Jack Copeland, *The Essential Turing*, Oxford University Press, 2004.
- Vladimir Tasic, *Mathematics and the Roots of Postmodern Thought*, Oxford University Press, 2001.
- Emile Borel, *Les nombres inaccessibles*, Gauthier-Villars, 1952.

- Lee Smolin, *Three Roads to Quantum Gravity,* Basic Books, 2001.

- Hans Christian von Baeyer, *Information,* Harvard University Press, 2004.

- Ernst Mayr, *This is Biology,* Harvard University Press, 1998.

- J. Wheeler, "It from bit," *Sakharov Memorial Lectures on Physics,* vol. 2, Nova Science, 1992.

- A. Zeilinger, "A foundational principle for quantum mechanics," *Found. Phys.* **29**, 631–643 (1999).

- G. 't Hooft, "The holographic principle," `http://arxiv.org/hep-th/0003004`.

- S. Lloyd, "The computational universe," `http://arxiv.org/quant-ph/0501135`.

- P. Zizzi, "A minimal model for quantum gravity," `http://arxiv.org/gr-qc/0409069`.

- J. Mäkelä, "Accelerating observers, area and entropy," `http://arxiv.org/gr-qc/0506087`.

- T. Jacobson, "Thermodynamics of spacetime," `http://arxiv.org/gr-qc/9504004`.

Is incompleteness
a serious problem?

Lecture at a meeting in Turin celebrating Gödel's 100th birthday.

In 1931 Kurt Gödel astonished the mathematical world by showing that no finite set of axioms can suffice to capture all of mathematical truth. He did this by constructing an assertion G_F about the whole numbers that manages to assert that it itself is unprovable (from a given finite set F of axioms using formal logic).[1]

G_F: "G_F cannot be proved from the finite set of axioms F."

This assertion G_F is therefore true if and only if it is unprovable, and the formal axiomatic system F in question either proves falsehoods (because it enables us to prove G_F) or fails to prove a true assertion (because it does not enable us to prove G_F). If we assume that the former situation is impossible, we conclude that F is necessarily incomplete since it does not permit us to establish the true statement G_F.

Either G_F is provable and F proves false statements,
or G_F is unprovable and therefore true, and F is incomplete.

Today, a century after Gödel's birth, the full implications of this "incompleteness" result are still quite controversial.[2]

---

[1] Gödel's paper is included in the well-known anthology [1].
[2] Compare for example the attitude in Franzén [2,3] with that in Chaitin [4,5,6].

An important step forward was achieved by Alan Turing in 1936. He showed that incompleteness could be derived as a corollary of uncomputability. Because if there are things that cannot be computed (Turing's halting problem), then these things also cannot be proven. More precisely, if there were a finite set of axioms F that always enabled us to prove whether particular programs P halt or fail to halt, then we could calculate whether a given program P halts or not by running through the tree of all possible deductions from the axioms F until we either find a proof that P halts or we find a proof that P never halts. But, as Turing showed in his famous 1936 paper "On Computable Numbers with an Application to the *Entscheidungsproblem*," there cannot be an algorithm for deciding whether or not individual programs P halt.[3]

If we can always prove whether or not P halts,
then we can always calculate whether or not P halts
(by systematically running through the tree of all possible proofs).

Now let's combine Turing's approach with ideas from Sections V and VI of Leibniz's *Discours de métaphysique* (1686). Consider the following toy model of what physicists do:

Theory **(program)** → *COMPUTER* → *Experimental Data* **(output).**

In other words, this is a software model of science, in which theories are considered to be programs for computing experimental data. In this toy model, the statement that the simplest theory is best corresponds to choosing the smallest, the most concise program for calculating the facts that we are trying to explain. And a key insight of Leibniz [7] is that if we allow arbitrarily complicated theories then the concept of theory becomes vacuous because there is always a theory. More precisely, in our software model for science this corresponds to the observation that if we have N bits of experimental data then our theory must be a program that is much less than N bits in size, because if the theory is allowed to have as many bits as the data, then there is always a theory.

Understanding = Compression!

Now let's abstract from this the concept of an "elegant" program:

---

[3]Turing's paper is also included in the collection [1].

**P is an elegant program if and only if
no smaller program Q written
in the same programming language
produces exactly the same output that P does.**

In our software model for science, the best theory is always an elegant program. Furthermore, there are infinitely many elegant programs, since for any computational task there is always at least one elegant program, and there are infinitely many computational tasks. However, what if we want to **prove** that a particular program P is elegant? Astonishingly enough, any finite set of axioms F can only enable us to prove that **finitely** many individual programs P are elegant!

Why is this the case? Consider the following paradoxical program P_F:

**P_F: The output of P_F is the same as
the output of the first provably elegant program Q
that is larger than P_F is.**

P_F runs through the tree of all possible deductions from the finite set of axioms F until it finds the first provably elegant program Q that is larger than P_F is, and then P_F simulates the computation that Q performs and then produces as its output the same output that Q produces. But this is impossible because P_F is too small to be able to produce that output! Assuming that F cannot enable us to prove false theorems, we must conclude that Q cannot exist. Thus if Q is an elegant program that is larger than P_F is, then the axioms F cannot enable us to prove that Q is elegant. Therefore F can only enable us to prove that **finitely** many individual programs Q are elegant. ***Q.E.D.***[4]

My personal belief, which is not shared by many in the mathematics community, is that modern incompleteness results such as this one push us in the direction of a "quasi-empirical" view of mathematics, in which we should be willing to accept new mathematical axioms that are not at all

---

[4]An immediate corollary is that the halting problem is unsolvable. For if we could determine all the programs that halt, then by running them and seeing their output we could also determine all the elegant programs, which we have just shown to be impossible. This program-size complexity argument for deriving the unsolvability of the halting problem is completely different from Turing's original 1936 proof, which is basically just an instance of Cantor's diagonal argument—from set theory—applied to the set of all computable real numbers.

self-evident but that are justified pragmatically, because they enable us to explain vast tracts of mathematical results. In other words, I believe that in mathematics, just as in physics, the function of theories is to enable us to compress many observations into a much more compact set of assumptions.[5]

So, in my opinion, incompleteness is extremely serious: It forces us to realize that *perhaps mathematics and physics are not as different as most people think.*[6]

Mathematics ≈ Physics?!

References

[1] Martin Davis, *The Undecidable,* Dover Publications, Mineola, New York, 2004.

[2] Torkel Franzén, *Gödel's Theorem,* A. K. Peters, Wellesley, Massachusetts, 2005.

[3] Torkel Franzén, "The popular impact of Gödel's incompleteness theorem," *Notices of the AMS,* April 2006, pp. 440–443.

[4] Gregory Chaitin, *Meta Math!,* Pantheon Books, New York, 2005.

[5] Gregory Chaitin, "The limits of reason," *Scientific American,* March 2006, pp. 74–81.[7]

[6] Gregory Chaitin, *Teoria algoritmica della complessità,* Giappichelli Editore, Turin, 2006.

[7] G. W. Leibniz, *Discours de métaphysique,* Gallimard, Paris, 1995.

[8] Thomas Tymoczko, *New Directions in the Philosophy of Mathematics,* Princeton University Press, Princeton, New Jersey, 1998.

[9] Jonathan Borwein and David Bailey, *Mathematics by Experiment,* A. K. Peters, Wellesley, Massachusetts, 2004.

---

[5] "Quasi-empirical" is a term invented by Lakatos. See Tymoczko [8].

[6] In this connection, see Borwein and Bailey [9] on the use of experimental methods in mathematics.

[7] Italian translation published in the May 2006 issue of *Le Scienze.*

Speculations on biology, information & complexity

It would be nice to have a mathematical understanding of basic biological concepts and to be able to prove that life must evolve in very general circumstances. At present we are far from being able to do this. But I'll discuss some partial steps in this direction plus what I regard as a possible future line of attack.

Can Darwinian evolution be made into a mathematical theory? Is there a fundamental mathematical theory for biology?

Darwin = math ?!

In 1960 the physicist Eugene Wigner published a paper with a wonderful title, "The unreasonable effectiveness of mathematics in the natural sciences." In this paper he marveled at the miracle that pure mathematics is so often extremely useful in theoretical physics.

To me this does not seem so marvelous, since mathematics and physics co-evolved. That however does not diminish the miracle that at a fundamental level Nature is ruled by simple, beautiful mathematical laws, that is, the miracle that Nature is comprehensible.

I personally am much more disturbed by another phenomenon, pointed out by I. M. Gel'fand and propagated by Vladimir Arnold in a lecture of his that is available on the web, which is the stunning contrast between the relevance of mathematics to physics, and its amazing lack of relevance to biology!

Indeed, unlike physics, biology is not ruled by simple laws. There is no equation for your spouse, or for a human society or a natural ecology. Biology is the domain of the complex. It takes 3×10^9 bases $= 6 \times 10^9$ bits of information to specify the DNA that determines a human being.

Darwinian evolution has acquired the status of a dogma, but to me as a mathematician seems woefully vague and unsatisfactory. What is evolution? What is evolving? How can we measure that? And can we **prove**, mathematically prove, that with high probability life must arise and evolve?

In my opinion, if Darwin's theory is as simple, fundamental and basic as its adherents believe, then there ought to be an equally fundamental mathematical theory about this, that expresses these ideas with the generality, precision and degree of abstractness that we are accustomed to demand in pure mathematics.

Look around you. We are surrounded by evolving organisms, they're everywhere, and their ubiquity is a challenge to the mathematical way of thinking. Evolution is not just a story for children fascinated by dinosaurs. In my own lifetime I have seen the ease with which microbes evolve immunity to antibiotics. We may well live in a future in which people will again die of simple infections that we were once briefly able to control.

Evolution seems to work remarkably well all around us, but not as a mathematical theory!

In the next section of this paper I will speculate about possible directions for modeling evolution mathematically. I do not know how to solve this difficult problem; new ideas are needed. But later in the paper I will have the pleasure of describing a minor triumph. The program-size complexity viewpoint that I will now describe to you does have some successes to its credit, even though they only take us an infinitesimal distance in the direction we must travel to fully understand evolution.

A software view of biology: Can we model evolution via evolving software?

I'd like to start by explaining my overall point of view. It is summarized here:

Life = Software ?

$$
\begin{array}{c}
\text{program} \longrightarrow COMPUTER \longrightarrow \text{output} \\
\text{DNA} \longrightarrow DEVELOPMENT/PREGNANCY \longrightarrow \text{organism}
\end{array}
$$

(Size of program in bits) $\approx$ (Amount of DNA in bases) $\times$ 2

So the idea is firstly that I regard life as software, biochemical software. In particular, I focus on the digital information contained in DNA. In my opinion, DNA is essentially a programming language for building an organism and then running that organism.

More precisely, my central metaphor is that DNA is a computer program, and its output is the organism. And how can we measure the complexity of an organism? How can we measure the amount of information that is contained in DNA? Well, each of the successive bases in a DNA strand is just 2 bits of digital software, since there are four possible bases. The alphabet for computer software is 0 and 1. The alphabet of life is A, G, C, and T, standing for adenine, cytosine, guanine, and thymine. A program is just a string of bits, and the human genome is just a string of bases. So in both cases we are looking at digital information.

My basic approach is to measure the complexity of a digital object by the size in bits of the smallest program for calculating it. I think this is more or less analogous to measuring the complexity of a biological organism by 2 times the number of bases in its DNA.

Of course, this is a tremendous oversimplification. But I am only searching for a toy model of biology that is simple enough that I can prove some theorems, not for a detailed theory describing the actual biological organisms that we have here on earth. I am searching for the Platonic essence of biology; I am only interested in the actual creatures we know and love to the extent that they are clues for finding ideal Platonic forms of life.

How to go about doing this, I am not sure. But I have some suggestions.

It might be interesting, I think, to attempt to discover a toy model for evolution consisting of evolving, competing, interacting programs. Each organism would consist of a single program, and we would measure its complexity in bits of software. The only problem is how to make the programs interact! This kind of model has no geometry, it leaves out the physical universe in which the organisms live. In fact, it omits bodies and retains only their DNA. This hopefully helps to make the mathematics more tractable. But at present this model has no interaction between organisms, no notion of time, no dynamics, and no reason for things to evolve. The question is how to add that to the model.

Hopeless, you may say. Perhaps not! Let's consider some other models that people have proposed. In von Neumann's original model creatures are embedded in a cellular automata world and are largely immobile. Not so good! There is also the problem of dissecting out the individual organisms that are embedded in a toy universe, which must be done before their individual complexities can be measured. My suggestion in one of my early papers that it might be possible to use the concept of mutual information—the extent to which the complexity of two things taken together is smaller than the sum of their individual complexities—in order to accomplish this, is not, in my current opinion, particularly fruitful.

In von Neumann's original model we have the complete physics for a toy cellular automata universe. Walter Fontana's ALChemy = algorithmic chemistry project went to a slightly higher level of abstraction. It used LISP S-expressions to model biochemistry. LISP is a functional programming language in which everything—programs as well as data—is kept in identical symbolic form, namely as what are called LISP S-expressions. Such programs can easily operate on each other and produce other programs, much in the way that molecules can react and produce other molecules.

I have a feeling that both von Neumann's cellular automata world and Fontana's algorithmic chemistry are too low-level to model biological evolution.[1] So instead I am proposing a model in which individual creatures are programs. As I said, the only problem is how to model the ecology in which these creatures compete. In other words, the problem is how to insert

---

[1] A model with perhaps the opposite problem of being at too **high** a level, is Douglas Lenat's AM = Automated Mathematician project, which dealt with the evolution of new mathematical concepts.

a dynamics into this static software world.[2]

Since I have not been able to come up with a suitable dynamics for the software model I am proposing, I must leave this as a challenge for the future and proceed to describe a few biologically relevant things that I **can** do by measuring the size of computer programs. Let me tell you what this viewpoint can buy us that is a tiny bit biologically relevant.

Pure mathematics has infinite complexity and is therefore like biology

Okay, program-size complexity can't help us very much with biological complexity and evolution, at least not yet. It's not much help in biology. But this viewpoint has been developed into a mathematical theory of complexity that I find beautiful and compelling—since I'm one of the people who created it—and that has important applications in another major field, namely metamathematics. I call my theory *algorithmic information theory,* and in it you measure the complexity of something X via the size in bits of the smallest program for calculating X, while completely ignoring the amount of effort which may be necessary to discover this program or to actually run it (time and storage space). In fact, we pay a severe price for ignoring the time a program takes to run and concentrating only on its size. We get a beautiful theory, but we can almost never be sure that we have found the smallest program for calculating something. We can almost never determine the complexity of anything, if we chose to measure that in terms of the size of the smallest program for calculating it!

This amazing fact, a modern example of the incompleteness phenomenon first discovered by Kurt Gödel in 1931, severely limits the practical utility of the concept of program-size complexity. However, from a philosophical point of view, this paradoxical limitation on what we can know is precisely the most interesting thing about algorithmic information theory, because that has profound epistemological implications.

---

[2]Thomas Ray's *Tierra* project did in fact create an ecology with software parasites and hyperparasites. The software creatures he considered were sequences of machine language instructions coexisting in the memory of a single computer and competing for that machine's memory and execution time. Again, I feel this model was too low-level. I feel that too much micro-structure was included.

The jewel in the crown of algorithmic information theory is the halting probability Ω, which provides a concentrated version of Alan Turing's 1936 halting problem. In 1936 Turing asked if there was a way to determine whether or not individual self-contained computer programs will eventually stop. And his answer, surprisingly enough, is that this cannot be done. Perhaps it can be done in individual cases, but Turing showed that there could be no general-purpose algorithm for doing this, one that would work for all possible programs.

The halting probability Ω is defined to be the probability that a program that is chosen at random, that is, one that is generated by coin tossing, will eventually halt. If no program ever halted, the value of Ω would be zero. If all programs were to halt, the value of Ω would be one. And since in actual fact some programs halt and some fail to halt, the value of Ω is greater than zero and less than one. Moreover, Ω has the remarkable property that its numerical value is maximally unknowable. More precisely, let's imagine writing the value of Ω out in binary, in base-two notation. That would consist of a binary point followed by an infinite stream of bits. It turns out that these bits are irreducible, both computationally and logically:

- You need an N-bit program in order to be able to calculate the first N bits of the numerical value of Ω.

- You need N bits of axioms in order to be able to prove what are the first N bits of Ω.

- In fact, you need N bits of axioms in order to be able to determine the positions and values of any N bits of Ω, not just the first N bits.

Thus the bits of Ω are, in a sense, mathematical facts that are *true for no reason,* more precisely, for no reason simpler than themselves. Essentially the only way to determine the values of some of these bits is to directly add that information as a new axiom.

And the only way to calculate individual bits of Ω is to separately add each bit you want to your program. The more bits you want, the larger your program must become, so the program doesn't really help you very much. You see, you can only calculate bits of Ω if you already **know** what these bits are, which is not terribly useful. Whereas with $\pi = 3.1415926\ldots$ we can get **all** the bits or all the digits from a single finite program, that's all you have to know. The algorithm for π compresses an infinite amount of information into a finite package. But with Ω there can be no compression, none at all, because there is absolutely no structure.

Furthermore, since the bits of Ω in their totality are infinitely complex, we see that pure mathematics contains infinite complexity. Each of the bits of Ω is, so to speak, a complete surprise, an individual atom of mathematical creativity. Pure mathematics is therefore, fundamentally, much more similar to biology, the domain of the complex, than it is to physics, where there is still hope of someday finding a theory of everything, a complete set of equations for the universe that might even fit on a T-shirt.

In my opinion, establishing this surprising fact has been the most important achievement of algorithmic information theory, even though it is actually a rather weak link between pure mathematics and biology. But I think it's an actual link, perhaps the first.

Computing Ω in the limit from below as a model for evolution

I should also point out that Ω provides an extremely abstract—much too abstract to be satisfying—model for evolution. Because even though Ω contains infinite complexity, it can be obtained in the limit of infinite time via a computational process. Since this extremely lengthy computational process generates something of infinite complexity, it may be regarded as an evolutionary process.

How can we do this? Well, it's actually quite simple. Even though, as I have said, Ω is maximally unknowable, there is a simple but very time-consuming way to obtain increasingly accurate lower bounds on Ω. To do this simply pick a cut-off t, and consider the finite set of all programs p up to t bits in size which halt within time t. Each such program p contributes $1/2^{|p|}$, 1 over 2 raised to p's size in bits, to Ω. In other words,

$$\Omega \;=\; \lim_{t \longrightarrow \infty} \left(\sum_{|p| \,\leq\, t \ \& \ \text{halts within time } t} 2^{-|p|} \right).$$

This may be cute, and I feel compelled to tell you about it, but I certainly do not regard this as a satisfactory model for biological evolution, since there is no apparent connection with Darwin's theory.

References

The classical work on a theoretical mathematical underpinning for biology is von Neumann's posthumous book [2].[3] Interestingly enough, Francis Crick—who probably contributed more than any other individual to creating modern molecular biology—for many years shared an office with Sydney Brenner, who was aware of von Neumann's thoughts on theoretical biology and self-reproduction. This interesting fact is revealed in the splendid biography of Crick [3].

For a book-length presentation of my own work on information and complexity, see [4], where there is a substantial amount of material on molecular biology. This book is summarized in my recent article [5], which however does not discuss biology. A longer overview of [4] is my Alan Turing lecture [6], which does touch on biological questions.

For my complete train of thought on biology extending over nearly four decades, see also [7,8,9,10,11].

For information on *Tierra,* see Tom Ray's home page at http://www.his.atr.jp/~ray/. For information on ALChemy, see http://www.santafe.edu/~walter/AlChemy/papers.html. For information on Douglas Lenat's Automated Mathematician, see [12] and the Wikipedia entry http://en.wikipedia.org/wiki/Automated_Mathematician.

For Vladimir Arnold's provocative lecture, the one in which Wigner and Gel'fand are mentioned, see http://pauli.uni-muenster.de/~munsteg/arnold.html. Wigner's entire paper is itself on the web at http://www.dartmouth.edu/~matc/MathDrama/reading/Wigner.html.

[1] J. Kemeny, "Man viewed as a machine," *Scientific American,* April 1955, pp. 58–67.

[2] J. von Neumann, *Theory of Self-Reproducing Automata,* University of Illinois Press, Urbana, 1967.

[3] M. Ridley, *Francis Crick,* Eminent Lives, New York, 2006.

[4] G. Chaitin, *Meta Math!,* Pantheon Books, New York, 2005.

[5] G. Chaitin, "The limits of reason," *Scientific American,* March 2006, pp. 74–81.

[6] G. Chaitin, "Epistemology as information theory: From Leibniz to Ω," European Computing and Philosophy Conference, Västeraas, Sweden, June 2005.

[7] G. Chaitin, "To a mathematical definition of 'life'," *ACM SICACT News,* January 1970, pp. 12–18.

[8] G. Chaitin, "Toward a mathematical definition of 'life'," R. Levine, M. Tribus, *The Maximum Entropy Formalism,* MIT Press, 1979, pp. 477–498.

[9] G. Chaitin, "Algorithmic information and evolution," O. Solbrig, G. Nicolis, *Perspectives on Biological Complexity,* IUBS Press, 1991, pp. 51–60.

[10] G. Chaitin, "Complexity and biology," *New Scientist,* 5 October 1991, p. 52.

---

[3]An earlier account of von Neumann's thinking on this subject was published in [1], which I read as a child.

[11] G. Chaitin, "Meta-mathematics and the foundations of mathematics," *Bulletin of the European Association for Theoretical Computer Science,* June 2002, pp. 167–179.

[12] D. Lenat, "Automated theory formation in mathematics," pp. 833–842 in volume 2 of R. Reddy, *Proceedings of the 5th International Joint Conference on Artificial Intelligence, Cambridge, MA, August 1977,* William Kaufmann, 1977.

How much information can there be in a real number?

This note gives some information about the magical number Ω and why it is of interest. Our purpose is to explain the significance of recent work by Calude and Dinneen attempting to compute Ω. Furthermore, we propose measuring human intellectual progress (not scientific progress) via the number of bits of Ω that can be determined at any given moment in time using the current mathematical theories.

1. Introduction

A real number corresponds to the length of a line segment that is measured with infinite precision. A rational number has a periodic decimal expansion. For example,

$$\frac{1}{3} = 0.3333333\ldots$$

The decimal expansion of an irrational real number is not periodic. Here are three well-known irrational reals that everyone encounters in high school and college mathematics: $\sqrt{2}$, π, and e.

Each of these numbers would seem to contain an infinite amount of information, because they have an infinite decimal expansion that never repeats. For example,

$$\pi = 3.1415926\ldots$$

However, π actually only contains a finite amount of information, because there is a small computer program for computing π. Instead of sending someone the digits of π, we can just explain to them how to compute as many digits as they want.

Are there any real numbers that contain an infinite amount of information? Well, clearly, if the successive decimal digits are chosen at random, the resulting stream of digits has no structure, each digit is a complete surprise, and there cannot be an algorithm for computing the number digit by digit.

However, this random sequence of digits is not **useful** information, not at all. It's an infinite amount of completely useless information.

2. Borel's Know-It-All Real Number

In 1927, the French mathematician Emile Borel pointed out that there are real numbers which contain an infinite amount of extremely **useful** information. The particular example that he gave is defined like this: Its Nth digit answers the Nth yes/no question in an infinite list of all possible yes/no questions, questions about the weather, the stock market, history, the future, physics, mathematics... Here I am talking about the Nth digit after the decimal point. Borel's number is between zero and one; there is nothing before the decimal point, only stuff after the decimal point. And we can assemble this list of questions because the set of all possible questions is what mathematicians call a countable or a denumerable set.

3. Using a Real Number as an Oracle for the Halting Problem

Borel's real number may seem rather unreal, rather fantastic, even though it exists in some Platonic, ideal, conceptual sense. How about a more realistic example, and now let's use base two, not base ten. Well, there is a real Θ whose Nth bit tells us whether or not the Nth computer program ever halts. This time we imagine an infinite list of all possible self-contained computer programs—not yes/no questions—and ask which programs will eventually finish running. This is Alan Turing's famous 1936 halting problem.

Θ doesn't tell us anything about the stock market or history, but it does tell us a great deal about mathematics. Why? Because knowing this number Θ would automatically enable us to resolve famous mathematical problems like Fermat's so-called last theorem, which asserts that there are no positive integer solutions for

$$x^N + y^N = z^N$$

with the power N greater than two.

How can Θ enable us to decide if Fermat was right and this equation has no solutions? There is a simple computer program for systematically searching for a solution of Fermat's equation. This program will fail to halt precisely if Fermat's conjecture that there are no solutions is correct.

However, in the case of Fermat's conjecture there is no need to wait for the number Θ; Andrew Wiles now has a proof that there are no solutions. But Θ would enable us to answer an infinite number of such conjectures, more precisely, all conjectures that can be refuted by a single counter example that we can search for using a computer.

4. N Cases of the Halting Problem is Only $\log_2 N$ Bits of Information

So knowing the answers to individual cases of the halting problem can be valuable information, and Θ enables us to answer **all** such problems, but unfortunately not in an optimal way. Θ isn't optimal, it is highly redundant, we're wasting lots of bits. Individual answers to the halting problem aren't independent, they're highly correlated.

Why? Because if we are given N programs, we can determine which ones halt and which ones don't if we merely know **how many** of these N programs halt, and to know that is only about $\log_2 N$ bits of information. (Run all N programs in parallel until precisely the correct number have stopped; the remaining programs will never stop.)

Furthermore, $\log_2 N$ is much smaller than N for all sufficiently large values of N.

So what is the best we can do? Is there an oracle for the halting problem that isn't redundant, that doesn't waste any bits?

5. The Halting Probability Ω is the Most Compact Oracle for the Halting Problem

The best way to pack information about the halting problem into a real number is to know a great many bits of the numerical value of the probability that a program chosen at random will eventually halt. Precisely how do I define this halting probability? Well, the exact definition is a little complicated, and in fact the numerical value of Ω depends on the particular computer and the programming language that you pick.

The general idea is that the computer that we are using flips a fair coin to generate each bit of the program, a heads yields a 1, a tails yields a 0, successive coin tosses are independent, and the computer starts running the program right away as it generates these bits. Ω is the probability that this process will eventually halt.

More precisely, each K-bit program p that halts contributes precisely $1/2^K$ to the halting probability Ω:

$$\Omega = \sum_{p \text{ halts}} 2^{-(\text{the size of } p \text{ in bits})}.$$

Furthermore, to avoid having this sum diverge to infinity, the set of meaningful programs must be a prefix-free set, in other words, no extension of a valid program is a valid program. Then what information theorists call the Kraft inequality applies to the set of all programs and Ω is necessarily less than one.

Ω is a very valuable oracle, because knowing the first N bits of Ω would enable us to resolve the halting problem for all programs up to N bits in size. No oracle for the halting problem can do better than this. Ω is so valuable precisely because it is the most compact way to represent this information. It's the best possible oracle for the halting problem. You get the biggest bang for your buck with each bit!

And because this information is so valuable, Ω is maximally unknowable, maximally uncomputable: An N-bit computer program can compute at most N bits of Ω, and a mathematical theory with N bits of axioms can enable us to determine at most N bits of Ω. In other words, the bits of Ω are incompressible, irreducible information, both logically irreducible and computationally irreducible.

Paradoxically, however, even though Ω is packed full of useful information, its successive bits **appear** to be totally unstructured and random, totally chaotic, because otherwise Ω would not be the most compact oracle for the halting problem. If one could predict future bits from past bits, then Ω would not be the best possible compression of all the answers to individual cases of Turing's halting problem.

6. Measuring Mathematical or Human Intellectual Progress in Terms of Bits of Ω

Counting how many bits of Ω our current mathematical theories permit us to know, gives us a way to measure the complexity of our mathematical knowledge as a function of time. Ω is infinitely complex, and at any given moment our theories capture at most a finite amount of this complexity. Our minds are finite, not infinitely complex like Ω.

But what if we bravely try to compute Ω anyway?

7. Storming the Heavens: Attempting to Compute the Uncomputable Bits of Ω

This amounts to a systematic attempt to increase the complexity of our mathematical knowledge, and it is precisely what Calude and Dinneen try to do in [1]. As they show, you can start off well enough and indeed determine a few of the initial bits of Ω. But as I have tried to explain, the further you go, the more creativity, the more ingenuity is required. To continue making progress, you will eventually need to come up with more and more complicated mathematical principles, novel principles that are not consequences of our current mathematical knowledge.

Will mathematics always be able to advance in this way, or will we eventually hit an insurmountable obstacle? Who knows! What is clear is that Ω can never be known in its entirety, but if the growth of our mathematical knowledge continues unabated, each individual bit of Ω can eventually be known.

I hope that this note gives some idea why [1] is of interest. (See also [2].) For more on Ω, please see my article in *Scientific American* [3] or my book [4]. A more recent paper is my Enriques lecture at the University of Milan in 2006 [5].

References

[1] C. S. Calude, M. J. Dinneen, "Exact approximations of Ω numbers," *International Journal of Bifurcation and Chaos* **17** (2007), in press.

[2] C. S. Calude, E. Calude, M. J. Dinneen, "A new measure of the difficulty of problems," *Journal of Multiple-Valued Logic and Soft Computing* **12** (2006), pp. 285–307.

[3] G. Chaitin, "The limits of reason," *Scientific American* **294**, No. 3 (March 2006), pp. 74–81.

[4] G. Chaitin, *Meta Math!*, Pantheon, New York, 2005, *Meta Maths*, Atlantic Books, London, 2006.

[5] G. Chaitin, "The halting probability Ω: Irreducible complexity in pure mathematics," *Milan Journal of Mathematics* **75** (2007), in press.

The halting probability Ω: Irreducible complexity in pure mathematics

Some Gödel centenary reflections on whether incompleteness is really serious, and whether mathematics should be done somewhat differently, based on using algorithmic complexity measured in bits of information. [Enriques lecture given Monday, October 30, 2006, at the University of Milan.]

Introduction: What is mathematics?

It is a pleasure for me to be here today giving this talk in a lecture series in honor of Frederigo Enriques. Enriques was a great believer in mathematical intuition, and disdained formal proofs. The work of Gödel, Turing and myself that I will review goes some way to justifying Enriques's belief in intuition. And, as you will see, I also agree with Enriques's emphasis on the importance of the philosophy and the history of science and mathematics.

This year is the centenary of Kurt Gödel's birth. Nevertheless, his famous 1931 incompleteness theorem remains controversial. To postmodernists, it justifies the belief that truth is a social construct, not absolute. Most mathematicians ignore incompleteness, and carry on as before, in a formalist, axiomatic, Hilbertian, Bourbaki spirit. I, on the contrary, have bet my life on the hunch that incompleteness is really serious, that it cannot be ignored, and that it means that mathematics is actually somewhat different from what most people think it is.

Gödel himself did not think that his theorem showed that mathematics has limitations. In several essays he made it clear that he believed that mathematicians could eventually settle any significant question by using their mathematical intuition, their ability to directly perceive the Platonic world of mathematical ideas, and by inventing or discovering new concepts and new axioms, new principles.

Furthermore, I share Enriques's faith in intuition. I think that excessive formalism and abstraction is killing mathematics. In my opinion math papers shouldn't attempt to replace all words by formulas, instead they should be like literary essays, they should attempt to explain and convince.

So let me tell you the story of metamathematics, of how mathematicians have tried to use mathematical methods to study the power and the limitations of math itself. It's a fairly dramatic story; in a previous era it might have been the subject of epic poems, of Iliads and Odysseys of verse. I'll start with David Hilbert.

Hilbert: Can mathematics be entombed in a formal axiomatic theory?

Hilbert stated the traditional belief that mathematics can provide absolute truth, complete certainty, that mathematical truth is black or white with no uncertainty. His contribution was to realize, to emphasize, that if this were the case, then there should be, there ought to be, a formal axiomatic theory, a theory of everything, for all of mathematics.

In practice, the closest we have come to this today is Zermelo-Fraenkel set theory with the axiom of choice, the formal theory ZFC using first-order logic, which seems to suffice for most contemporary mathematics.

Hilbert did not invent mathematical logic, he took advantage of work going back to Leibniz, de Morgan, Boole, Frege, Peano, Russell and Whitehead, etc. But in my opinion he enunciated more clearly than anyone before him the idea that if math provides absolute truth, complete certainty, then there should be a finite set of axioms that we can all agree on from which it would in principle be possible to prove all mathematical truths by mechanically following the rules of formal mathematical logic. It would be slow, but it would work like an army of reason marching inexorably forward. It would make math into a merciless machine.

Hilbert did not say that mathematics should actually be done in this extremely formal way in which proofs are broken down into their atomic steps, with nothing omitted, excruciatingly detailed, using symbolic logic instead of a normal human language. But the idea was to eliminate all uncertainty, to make clear exactly when a proof is valid, so that this can be checked mechanically, thus making mathematical truth completely objective, eliminating all subjective elements, all matters of opinion.

Hilbert started with Peano arithmetic, but his ultimate goal was to axiomatize analysis and then all of mathematics, absolutely everything. In 1931, however, Gödel surprised everyone by showing that it couldn't be done, it was impossible.

Gödel: "This statement is unprovable!"

In fact, Gödel showed that no finite set of axioms suffice for elementary number theory, for the theory of 0, 1, 2, ... and addition and multiplication, that is, for Peano arithmetic. His proof is very strange. First of all he numbers all possible assertions and all possible proofs in Peano arithmetic. This converts the assertion that x is a proof of y into an arithmetic assertion about x and y.

Next Gödel constructs an assertion that refers to itself indirectly. It says that if you calculate a certain number, that gives you the number of an unprovable assertion, and this is done in such a way that we get an arithmetic statement asserting its own unprovability.

Consider "I am unprovable." It is either provable or not. If provable, we are proving a false assertion, which we very much hope is impossible. The only alternative left is that "I'm unprovable" is unprovable. If so it is true but unprovable, and there is a hole in formal mathematics, a true assertion that we cannot prove. In other words, our formal axiomatic theory must be incomplete, if we assume that only true assertions can be proved, which we fervently hope to be the case. Proving false assertions is even worse than not being able to prove a true assertion!

So that's Gödel's famous 1931 incompleteness theorem, and it was a tremendous shock to everyone. When I was a young student I read essays by John von Neumann, Hermann Weyl and others attesting to what a shock it was. I didn't realize that the generation that cared about this had been swept away by the Second World War and that mathematicians were going on exactly as before, ignoring Gödel. In fact, I thought that what Gödel discovered was only the tip of the iceberg. I thought that the problem had to be really serious, really profound, and that the traditional philosophy of math couldn't be slightly wrong, so that even a small scratch would shatter it into pieces. It had to be all wrong, in my opinion.

Does that mean that if you cannot prove a result that you like and have some numerical evidence for, if you cannot do this in a week, then invoking Gödel, you just add this conjectured result as a new axiom?! No, not at all, that is too extreme a reaction. But, as I will explain later, I do have something like that in mind.

You see, the real problem with Gödel's proof is that it gives no idea how serious incompleteness is. Gödel's true but unprovable assertion is bizarre, so it is easy to shrug it off. But if it turns out that incompleteness is pervasive, is ubiquitous, that is another matter.

And an important first step in the direction of showing that incompleteness is really serious was taken only five years later, in 1936, by Alan Turing, in a famous paper "On computable numbers. . . "

Turing: Most real numbers are uncomputable!

This paper is remembered, in fact, celebrated nowadays, for proposing a toy model of the computer called a Turing machine, and for its discussion of what we now call the halting problem. But this paper is actually about distinguishing between computable and uncomputable real numbers, numbers like π or e or $\sqrt{2}$ that we can compute with infinite precision, with arbitrary accuracy, and those that we cannot.

Yes, it's true, Turing's paper does contain the idea of software as opposed to hardware, of a universal digital machine that can simulate any other special-purpose digital machine. Mathematicians refer to this as a universal Turing machine, and, as I learned from von Neumann, it is the conceptual basis for all computer technology. But even more interesting is the fact that it is easy to see that most real numbers are uncomputable, and the new perspective this gives on incompleteness, as Turing himself points out. Let me summarize his discussion.

First of all, all possible software, all possible algorithms, can be placed in an infinite list and numbered 1, 2, 3, ... and so this set is denumerable or countable. However, the set of real numbers is, as Cantor showed, a higher-order infinity, it's uncountable, nondenumerable. Therefore most real numbers must be uncomputable. In Turing's paper he exhibits a single example of an uncomputable real, one that is obtained by applying Cantor's diagonal method to the list of all computable real numbers to obtain a new and different real, one that is not in the list.

In fact, using ideas that go back to Emile Borel, it is easy to see that if you pick a real number between 0 and 1 at random, with uniform probability, it is possible to pick a computable real, but this is infinitely improbable, because the computable reals are a set of measure zero, they can be covered with intervals whose total length is arbitrarily small. Just cover the first computable real in [0,1] with an interval of size $\epsilon/2$, the second real with an interval of size $\epsilon/4$, and in general the Nth real with an interval of size $\epsilon/2^N$. This covering has lengths totalling exactly ϵ, which can be made as small as we want.

Turing does not, however, make this observation. Instead he points out that it looks easy to compute his uncomputable real by taking the Nth digit produced by the Nth program and changing it. Why doesn't this work? Because we can never decide if the Nth program will ever produce an Nth digit! If we could, we could actually diagonalize over all computable reals and calculate an uncomputable real, which is impossible. And being able to decide if the Nth program will ever output an Nth digit is a special case of Turing's famous halting problem.

Note that there is no problem if you place an upper bound on the time allowed for a computation. It is easy to decide if the Nth program outputs an Nth digit in a trillion years, all you need to do is be patient and try it and see. Turing's halting problem is only a problem if there is no time limit. In other words, this is a deep conceptual problem, not a practical limitation on what we can do.

And, as Turing himself points out, incompleteness is an immediate corollary. For let's say we'd like to be able to prove whether individual computer programs, those that are self-contained and read no input, eventually halt or not. There can be no formal axiomatic theory for this, because if there were, by systematically running through the tree of all possible proofs, all possible deductions from the axioms using formal logic, we could always eventually decide whether an individual program halts or not, which is impossible.

In my opinion this is a fundamental step forward in the philosophy of mathematics because it makes incompleteness seem much more concrete and much more natural. It's almost a problem in physics, it's about a machine, you just ask whether or not it's going to eventually stop, and it turns out there's no way, no general way, to answer that question.

Let me emphasize that if a program does halt, we can eventually discover that. The problem, an extremely deep one, is to show that a program will never halt if this is in fact so. One can settle many special cases, even an infinity of them, but no finite set of axioms can enable you to settle all possible cases.

My own work takes off from here. My approach to incompleteness follows Turing, not Gödel. Later I'll consider the halting probability Ω and show that this number is wildly, in fact maximally, uncomputable and unknowable. I'll take Turing's halting problem and convert it into a real number...

My approach is very 1930's. All I add to Turing is that I measure software complexity, I look at the size of computer programs. In a moment, I'll tell you how these ideas actually go back to Leibniz. But first, let me tell you more about Borel's ideas on uncomputable reals, which are closely related to Turing's ideas.

Borel: Know-it-all and unnameable reals

Borel in a sense anticipated Turing, because he came up with an example of an uncomputable real in a small paper published in 1927. Borel's idea was to use the digits of a single real number as an oracle that can answer any yes/no question. Just imagine a list of all possible yes/no questions in French, said Borel. This is obviously a countable infinity, so there is an Nth question, an $N + 1$th question, etc. And you can place all the answers in the decimal expansion of a single real number; just use the Nth digit to answer the Nth question. Questions about history, about math, about the stock market!

So, says Borel, this real number exists, but to him it is a mathematical fantasy, not something real. Basically Borel has a constructive attitude, he believes that something exists only if we can calculate it, and Borel's oracle number can certainly not be calculated.

Borel didn't linger over this, he made his point and moved on, but his example is in my opinion a very interesting one, and will later lead us step by step to my Ω number.

Before I used Borel's ideas on measure and probability to point out that Turing's computable reals have measure zero, they're infinitely unlikely. Borel however seemed unaware of Turing's work. His own version of these ideas, in his final book, written when he was 80, *Les nombres inaccessibles,* is to point out that the set of reals that can somehow be individually identified, constructively or not, has measure zero, because the set of all possible descriptions of a real is countable. Thus, with probability one, a real cannot be uniquely specified, it can never be named, there are simply not enough names to go around!

The real numbers are the simplest thing in the world geometrically, they are just points on a line. But arithmetically, as individuals, real numbers are actually rather unreal. Turing's 1936 uncomputable real is just the tip of the iceberg, the problem is a lot more serious than that.

Let me now talk about looking at the size of computer programs and what that has to tell us about incompleteness. To explain why program size is important, I have to start with Leibniz, with some ideas in his 1686 *Discours de métaphysique,* which was found among his papers long after his death.

Theories as software, Understanding as compression, Lawless incompressible facts

The basic model of what I call algorithmic information theory (AIT) is that a scientific theory is a computer program that enables you to compute or explain your experimental data:

$$\text{theory (program)} \longrightarrow \textbf{Computer} \longrightarrow \text{data (output)}.$$

In other words, the purpose of a theory is to compute facts. The key observation of Leibniz is that there is always a theory that is as complicated as the facts it is trying to explain. This is useless: a theory is of value only to the extent that it compresses a great many bits of data into a much smaller number of bits of theory.

In other words, as Hermann Weyl put it in 1932, the concept of law becomes vacuous if an arbitrarily complicated law is permitted, for then there is always a law. A law of nature has to be much simpler than the data it explains, otherwise it explains nothing. The problem, asks Weyl, is how can we measure complexity? Looking at the size of equations is not very satisfactory.

AIT does this by considering both theories and data to be digital information; both are a finite string of bits. Then it is easy to compare the size of the theory with the size of the data it supposedly explains, by merely counting the number of bits of information in the software for the theory and comparing this with the number of bits of experimental data that we are trying to understand.

Leibniz was actually trying to distinguish between a lawless world and one that is governed by law. He was trying to elucidate what it means to say that science works. This was at a time when modern science, then called mechanical philosophy, was just beginning; 1686 was the year before Leibniz's nemesis Newton published his *Principia*.

Leibniz's original formulation of these ideas was like this. Take a piece of paper, and spot it with a quill pen, so that you get a finite number of random points on a page. There is always a mathematical equation that passes precisely through these points. So this cannot enable you to distinguish between points that are chosen at random and points that obey a law. But if the equation is simple, then that's a law. If, on the contrary, there is no simple equation, then the points are lawless, random.

So part and parcel of these ideas is a definition of randomness or lawlessness for finite binary strings, as those which cannot be compressed into a program for calculating them that is substantially smaller in size. In fact, it is easy to see that most finite binary strings require programs of about the same size as they are. So these are the lawless, random or algorithmically irreducible strings, and they are the vast majority of all strings. Obeying a law is the exception, just as being able to name an individual real is an exception.

Let's go a bit further with this theories as software model. Clearly, the best theory is the simplest, the most concise, the smallest program that calculates your data. So let's abstract things a bit, and consider what I call **elegant** programs:

- A program is elegant if no smaller program written in the same language, produces the same output.

In other words, an elegant program is the optimal, the simplest theory for its output. How can we be sure that we have the best theory? How can we tell whether a program is elegant? The answer, surprisingly enough, is that we can't!

Provably elegant programs

To show this, consider the following paradoxical program P:

- P computes the output of the first provably elegant program larger than P.

In other words, P systematically deduces all the consequences of the axioms, which are all the theorems in our formal axiomatic theory. As it proves each theorem, P examines it. First of all, P filters out all proofs that do not show that a particular program is elegant. For example, if P finds a proof of the Riemann hypothesis, it throws that away; it only keeps proofs that programs are elegant. And as it proves that individual programs are elegant, it checks each provably elegant program to see if this program is larger than P. As soon as P finds a provably elegant program that is larger than it is, it starts running that program, and produces that program's output as its own output. In other words, P's output is precisely the same as the output of the first provably elegant program that is larger than P.

However, P is too small to produce the same output as an elegant program that is larger than P, because this contradicts the definition of elegance! What to do? How can we avoid the contradiction?

First of all, we are assuming that our formal axiomatic theory only proves true theorems, and in particular, that if it proves that a program is elegant, this is in fact the case. Furthermore, the program P is not difficult to write out; I've done this in one of my books using the programming language called LISP. So the only way out is if P never finds the program it is looking for! In other words, the only way out is if it is never possible to prove that a program that's larger than P is elegant! But there are infinitely many possible elegant programs, and they can be arbitrarily big. But provably elegant programs can't be arbitrarily big, they can't be larger than P.

So how large is P, that's the key question. Well, the bulk of P is actually concerned with systematically producing all the theorems in our formal axiomatic theory. So I'll define the complexity of a formal axiomatic theory to be the size in bits of the smallest program for doing that. Then we can restate our metatheorem like this: You can't prove that a program is elegant if its size in bits is substantially larger than the complexity of the formal axiomatic theory you are working with. In other words, using a formal axiomatic theory with N bits of complexity, you can't prove that any program larger than $N + c$ bits in size is elegant. Here the constant c is the size in bits of the main program in P, the fixed number of bits not in that big N-bit subroutine for running the formal axiomatic theory and producing all its theorems.

Loosely put:

- You need an N-bit theory to show that an N-bit program is elegant.[1]

Why is this so interesting? Well, right away it presents incompleteness in an entirely new light. How? Because it shows that mathematics has infinite complexity, but any formal axiomatic theory can only capture a finite part of this complexity. In fact, just knowing which programs are elegant has infinite complexity.

So this makes incompleteness very natural; math has an infinite basis, no finite basis will do. Now incompleteness is the most natural thing in the world; it's not at all mysterious!

Now let me tell you about the halting probability Ω, which shows even better that math is infinitely complex.

What is the halting probability Ω?

Let's start with Borel's know-it-all number, but now let's use the Nth binary digit to tell us whether or not the Nth computer program ever halts. So now Borel's number is an oracle for the halting problem. For example, there is a bit which tells us whether or not the Riemann hypothesis is true, for that is equivalent to the statement that a program that systematically searches for zeros of the zeta function that are in the wrong place, never halts.

It turns out that this number, which I'll call Turing's number even though it does not occur in Turing's paper, is wasting bits, it is actually highly redundant. We don't really need N bits to answer N cases of the halting problem, a much smaller number of bits will do. Why?

Well, consider some large number N of cases of the halting problem, some large number N of individual programs for which we want to know whether or not each one halts. Is this really N bits of mathematical information? No,

---

[1]By the way, it is an immediate corollary that the halting problem is unsolvable, because if we could decide which programs halt, then we could run all the programs that halt and see what they output, and this would give us a way to determine which halting programs are elegant, which we've just shown is impossible. This is a new information-theoretic proof of Turing's theorem, rather different from Turing's original diagonal-argument proof.

the answers are not independent, they are highly correlated. How? Well, in order to answer N cases of the halting problem, we don't really need to know each individual answer; it suffices to know how many of these N programs will eventually halt. Once we know this number, which is only about $\log_2 N$ bits of information, we can run the N programs in parallel until exactly this number of them halt, and then we know that none of the remaining programs will ever halt. And $\log_2 N$ is much, much less than N for all sufficiently large N. In other words, Turing's number isn't the best possible oracle for the halting problem. It is highly redundant, it uses far too many bits.

Using essentially this idea, we can get the best possible oracle number for the halting problem; that is the halting probability Ω, which has no redundancy, none at all.

I don't have time to explain this in detail, but here is a formula for the halting probability:

$$\Omega = \sum_{p \text{ halts}} 2^{-|p|}.$$

The idea is that each K-bit program that halts contributes exactly $1/2^K$ to the halting probability Ω. In other words, Ω is the halting probability of a program p whose bits are generated by independent tosses of a fair coin.

Technical point: For this to work, for this sum to converge to a number between 0 and 1 instead of diverging to infinity, it is important that programs be self-delimiting, that no extension of a valid program be a valid program. In other words, our computer must decide by itself when to stop reading the bits of the program without waiting for a blank endmarker. In old Shannon-style information theory this is a well-known lemma called the Kraft inequality that applies to prefix-free sets of strings, to sets of strings which are never prefixes or extensions of each other. And an extended, slightly more complicated, version of the Kraft inequality plays a fundamental role in AIT.

I should also point out that the precise numerical value of Ω depends on your choice of computer programming language, or, equivalently, on your choice of universal self-delimiting Turing machine. But its surprising properties do not, they hold for a large class of universal Turing machines.

Anyway, once you fix the programming language, the precise numerical value of Ω is determined, it's well-defined. Let's imagine having Ω written out in base-two binary notation:

$$\Omega = .110110\ldots$$

These bits are totally lawless, algorithmically irreducible mathematical facts. They cannot be compressed into any theory smaller than they are.

More precisely, the bits of the halting probability Ω are both computationally and logically irreducible:

- You need an N-bit program to calculate N bits of Ω (any N bits, not just the first N).

- You need an N-bit theory to be able to determine N bits of Ω (any N bits, not just the first N).

Ω is an extreme case of total lawlessness; in effect, it shows that God plays dice in pure mathematics. More precisely, the bits of Ω refute Leibniz's principle of sufficient reason, because they are mathematical facts that are true for no reason (no reason simpler than they are). Essentially the only way to determine bits of Ω is to directly add these bits to your axioms. But you can prove anything by adding it as a new axiom; that's not using reasoning!

Why does Ω have these remarkable properties? Well, because it's such a good oracle for the halting problem. In fact, knowing the first N bits of Ω enables you to answer the halting problem for all programs up to N bits in size. And you can't do any better; that's why these bits are incompressible, irreducible information. If you think of what I called Turing's number as a piece of coal, then Ω is the diamond that you get from this coal by subjecting it to very high temperatures and pressures. A relatively small number of bits of Ω would in principle enable you to tell whether or not the Riemann hypothesis is false.

Concluding discussion

So Ω shows us, directly and immediately, that math has infinite complexity, because the bits of Ω are infinitely complex. But any formal axiomatic theory only has a finite, in fact, a rather small complexity, otherwise we wouldn't believe in it! What to do? How can we get around this obstacle? Well, by increasing the complexity of our theories, by adding new axioms, complicated axioms that are pragmatically justified by their usefulness instead of simple self-evident axioms of the traditional kind.

Here are some recent examples:

- The hypothesis that $\mathbf{P} \neq \mathbf{NP}$ in theoretical computer science.

- The axiom of projective determinacy in abstract set theory.

- Various versions of the Riemann hypothesis in analytic number theory.

In other words, I am advocating a "quasi-empirical" view of mathematics, a term that was invented by Imre Lakatos, by the way. (He wouldn't necessarily approve of the way I'm using it, though.)

To put it bluntly, from the point of view of AIT, mathematics and physics are not that different. In both cases, theories are compressions of facts, in one case facts we discover in a physics lab, in the other case, numerical facts discovered using a computer. Or, as Vladimir Arnold so nicely puts it, math is like physics, except that the experiments are cheaper! I'm not saying that math and physics are the same, but I am saying that maybe they are not as different as most people think.

Another way to put all of this, is that the DNA for pure math, Ω, is infinitely complex, whereas the human genome is 3×10^9 bases $= 6 \times 10^9$ bits, a large number, but a finite one. So pure math is even more complex than the traditional domain of the complicated, biology! Math does not have finite complexity the way that Hilbert thought, not at all, on the contrary!

These are highly heretical suggestions, suggestions that the mathematics community is extremely uncomfortable with. And I have to confess that I have attempted to show that incompleteness is serious and that math should be done somewhat differently, but I haven't been able to make an absolutely watertight case. I've done the best I can with one lifetime of effort, though.

But if you really started complicating mathematics by adding new non-self-evident axioms, what would happen? Might mathematics break into separate factions? Might different groups with contradictory axioms go to war? Hilbert thought math was an army of reason marching inexorably forward, but this sounds more like anarchy!

Perhaps anarchy isn't so bad; it's better than a prison, and it leaves more room for intuition and creativity. I think that Enriques might have been sympathetic to this point of view. After all, as Cantor, who created a crazy, theological, paradoxical theory of infinite magnitudes, said, the essence of mathematics resides in its freedom, in the freedom to imagine and to create.

Bibliography

For more on this, please see my book [4] that has just been published in Italian, or my previous book, which is currently available in three separate editions [1, 2]. Some related books in Italian [3, 5] and papers [6–8] are also listed below. The classic papers of 1931 and 1936 by Gödel and Turing are reprinted in the superb anthology [9]. Borel's 1927 know-it-all number may be found on page 275 of [10].

[1] G. Chaitin, *Meta Math!*, Pantheon, New York, 2005 (hardcover), Vintage, New York, 2006 (softcover).

[2] G. Chaitin, *Meta Maths*, Atlantic Books, London, 2006.

[3] U. Pagallo, *Introduzione alla filosofia digitale. Da Leibniz a Chaitin*, Giappichelli, Turin, 2005.

[4] G. Chaitin, *Teoria algoritmica della complessità*, Giappichelli, Turin, 2006.

[5] U. Pagallo, *Teoria giuridica della complessità*, Giappichelli, Turin, 2006.

[6] G. Chaitin, "How real are real numbers?," *International Journal of Bifurcation and Chaos* **16** (2006), pp. 1841–1848.

[7] G. Chaitin, "Epistemology as information theory: From Leibniz to Ω," *Collapse* **1** (2006), pp. 27–51.

[8] G. Chaitin, "The limits of reason," *Scientific American* **294**, No. 3 (March 2006), pp. 74–81. (Also published as "I limiti della ragione," *Le Scienze*, May 2006, pp. 66–73.)

[9] M. Davis, *The Undecidable*, Raven Press, Hewlett, 1965, Dover, Mineola, 2004.

[10] E. Borel, *Leçons sur la théorie des fonctions*, Gauthier-Villars, Paris, 1950, Gabay, Paris, 2003.

The halting probability Ω: Concentrated creativity

The number Ω is the probability that a self-contained computer program chosen at random, a program whose bits are picked one by one by tossing a coin, will eventually stop, rather than continue calculating forever:

$$\Omega = \sum_{p \text{ halts}} 2^{-|p|}.$$

Surprisingly enough, the precise numerical value of Ω is uncomputable, in fact, irreducibly complex.

Ω can be interpreted pessimistically, as indicating there are limits to human knowledge. The optimistic interpretation, which I prefer, is that Ω shows that one cannot do mathematics mechanically and that intuition and creativity are essential. Indeed, in a sense Ω is the crystalized, concentrated essence of mathematical creativity. — *Gregory Chaitin*

[*This is my contribution to a collection of* **Formulas for the Twenty-First Century**, *each explained in 120 words or less, assembled by art curator Hans-Ulrich Obrist.*]

List of publications

- G. Chaitin, "An improvement on a theorem of E. F. Moore," *IEEE Transactions on Electronic Computers* **EC-14** (1965), pp. 466–467.

- G. Chaitin, "On the length of programs for computing finite binary sequences by bounded-transfer Turing machines," *AMS Notices* **13** (1966), p. 133.

- G. Chaitin, "On the length of programs for computing finite binary sequences by bounded-transfer Turing machines II," *AMS Notices* **13** (1966), pp. 228–229.

- G. Chaitin, "On the length of programs for computing finite binary sequences," *Journal of the ACM* **13** (1966), pp. 547–569.

- G. Chaitin, "On the length of programs for computing finite binary sequences: Statistical considerations," *Journal of the ACM* **16** (1969), pp. 145–159.

- G. Chaitin, "On the simplicity and speed of programs for computing infinite sets of natural numbers," *Journal of the ACM* **16** (1969), pp. 407–422.

- G. Chaitin, "On the difficulty of computations," *IEEE Transactions on Information Theory* **IT-16** (1970), pp. 5–9. Reprinted in G. Chaitin, *Thinking about Gödel & Turing,* World Scientific, 2007.

- G. Chaitin, "To a mathematical definition of 'life'," *ACM SICACT News,* No. 4 (Jan. 1970), pp. 12–18.

- G. Chaitin, "Computational complexity and Gödel's incompleteness theorem," *AMS Notices* **17** (1970), p. 672.

- G. Chaitin, "Computational complexity and Gödel's incompleteness theorem," *ACM SIGACT News,* No. 9 (Apr. 1971), pp. 11–12.

- G. Chaitin, "Information-theoretic aspects of the Turing degrees," *AMS Notices* **19** (1972), pp. A-601, A-602.

- G. Chaitin, "Information-theoretic aspects of Post's construction of a simple set," *AMS Notices* **19** (1972), p. A-712.

- G. Chaitin, "On the difficulty of generating all binary strings of complexity less than n," *AMS Notices* **19** (1972), p. A-764.

- G. Chaitin, "On the greatest natural number of definitional or information complexity $\leq n$," *Recursive Function Theory: Newsletter,* No. 4 (Jan. 1973), pp. 11–13.

- G. Chaitin, "A necessary and sufficient condition for an infinite binary string to be recursive," *Recursive Function Theory: Newsletter,* No. 4 (Jan. 1973), p. 13.

- G. Chaitin, "There are few minimal descriptions," *Recursive Function Theory: Newsletter,* No. 4 (Jan. 1973), p. 14.

- G. Chaitin, "Information-theoretic computational complexity," *Abstracts of Papers, 1973 IEEE International Symposium on Information Theory,* p. F1-1.

- G. Chaitin, "Information-theoretic computational complexity," *IEEE Transactions on Information Theory* **IT-20** (1974), pp. 10–15. Reprinted in T. Tymoczko, *New Directions in the Philosophy of Mathematics,* Birkhäuser, 1986, *Expanded Edition,* Princeton University Press, 1998. Reprinted in G. Chaitin, *Thinking about Gödel & Turing,* World Scientific, 2007.

- G. Chaitin, "Information-theoretic limitations of formal systems," *Journal of the ACM* **21** (1974), pp. 403–424.

- G. Chaitin, "A theory of program size formally identical to information theory," *Abstracts of Papers, 1974 IEEE International Symposium on Information Theory,* p. 2.

- G. Chaitin, "Randomness and mathematical proof," *Scientific American* **232**, No. 5 (May 1975), pp. 47–52. Reprinted in N. H. Gregersen, *From Complexity to Life,* Oxford University Press, 2003. Reprinted in G. Chaitin, *Thinking about Gödel & Turing,* World Scientific, 2007.

- G. Chaitin, "A theory of program size formally identical to information theory," *Journal of the ACM* **22** (1975), pp. 329–340.

- G. Chaitin, "Information-theoretic characterizations of recursive infinite strings," *Theoretical Computer Science* **2** (1976), pp. 45–48.

- G. Chaitin, "Algorithmic entropy of sets," *Computers & Mathematics with Applications* **2** (1976), pp. 233–245.

- G. Chaitin, "Program size, oracles, and the jump operation," *Osaka Journal of Mathematics* **14** (1977), pp. 139–149.

- G. Chaitin, "Algorithmic information theory," *IBM Journal of Research and Development* **21** (1977), pp. 350–359, 496.

- G. Chaitin, "Recent work on algorithmic information theory," *Abstracts of Papers, 1977 IEEE International Symposium on Information Theory,* p. 129.

- G. Chaitin, J. T. Schwartz, "A note on Monte Carlo primality tests and algorithmic information theory," *Communications on Pure and Applied Mathematics* **31** (1978), pp. 521–527.

- G. Chaitin, "Toward a mathematical definition of 'life'," in R. D. Levine, M. Tribus, *The Maximum Entropy Formalism,* MIT Press, 1979, pp. 477–498.

- G. Chaitin, "Algorithmic information theory," *Encyclopedia of Statistical Sciences,* Vol. 1, Wiley, 1982, pp. 38–41.

- G. Chaitin, "Gödel's theorem and information," *International Journal of Theoretical Physics* **21** (1982), pp. 941–954. Reprinted in T. Tymoczko, *New Directions in the Philosophy of Mathematics,* Birkhäuser, 1986, *Expanded Edition,* Princeton University Press, 1998. Reprinted in Polish in R. Murawski, *Współczesna Filozofia Matematyki,* PWN, 2002. Reprinted in G. Chaitin, *Thinking about Gödel & Turing,* World Scientific, 2007.

- G. Chaitin, "Randomness and Gödel's theorem," *Mondes en développement,* No. 54–55 (1986), pp. 125–128.

- G. Chaitin, "Incompleteness theorems for random reals," *Advances in Applied Mathematics* **8** (1987), pp. 119–146.

- G. Chaitin, *Algorithmic Information Theory,* Cambridge University Press, 1987.

- G. Chaitin, *Information, Randomness & Incompleteness,* World Scientific, 1987.

- G. Chaitin, "Computing the busy beaver function," in T. M. Cover, B. Gopinath, *Open Problems in Communication and Computation,* Springer-Verlag, 1987, pp. 108–112.

- G. Chaitin, "An algebraic equation for the halting probability," in R. Herken, *The Universal Turing Machine,* Oxford University Press, 1988, pp. 279–283.

- G. Chaitin, "Randomness in arithmetic," *Scientific American* **259**, No. 1 (Jul. 1988), pp. 80–85. Reprinted in G. Chaitin, *Thinking about Gödel & Turing,* World Scientific, 2007.

- G. Chaitin, *Algorithmic Information Theory,* 2nd printing (with revisions), Cambridge University Press, 1988.

- G. Chaitin, *Algorithmic Information Theory,* 3rd printing (with revisions), Cambridge University Press, 1990.

- G. Chaitin, "Algorithmic information & evolution," in O. T. Solbrig, G. Nicolis, *Perspectives on Biological Complexity,* IUBS Press, 1991, pp. 51–60.

- G. Chaitin, *Information, Randomness & Incompleteness,* 2nd edition, World Scientific, 1990. Contains all the preceding papers in this list.

- G. Chaitin, "A random walk in arithmetic," *New Scientist* **125**, No. 1709 (24 Mar. 1990), pp. 44–46. Reprinted in N. Hall, *The New Scientist Guide to Chaos,* Penguin, 1992, (UK edition), *Exploring Chaos,* Norton, 1993, (US edition).

- G. Chaitin, "Le hasard des nombres," *La Recherche* **22**, No. 232 (May 1991), pp. 610–615. Reprinted in *La Recherche,* Hors-Série No. 2 (Aug. 1999), pp. 60–65.

- G. Chaitin, "Complexity and biology," *New Scientist* **132**, No. 1789 (5 Oct. 1991), p. 52.

- G. Chaitin, "LISP program-size complexity," *Applied Mathematics and Computation* **49** (1992), pp. 79–93.

- G. Chaitin, "Information-theoretic incompleteness," *Applied Mathematics and Computation* **52** (1992), pp. 83–101.

- G. Chaitin, "LISP program-size complexity II," *Applied Mathematics and Computation* **52** (1992), pp. 103–126.

- G. Chaitin, "LISP program-size complexity III," *Applied Mathematics and Computation* **52** (1992), pp. 127–139.

- G. Chaitin, "LISP program-size complexity IV," *Applied Mathematics and Computation* **52** (1992), pp. 141–147.

- G. Chaitin, "A Diary on Information Theory," *The Mathematical Intelligencer* **14**, No. 4 (Fall 1992), pp. 69–71.

- G. Chaitin, *Information-Theoretic Incompleteness*, World Scientific, 1992. Includes the preceding nine papers.

- G. Chaitin, *Algorithmic Information Theory*, 4th printing, Cambridge University Press, 1992. Identical to 3rd printing.

- G. Chaitin, "Randomness in arithmetic and the decline and fall of reductionism in pure mathematics," *Bulletin of the European Association for Theoretical Computer Science* **50** (Jun. 1993), pp. 314–328. Reprinted in J. Cornwell, *Nature's Imagination*, Oxford University Press, 1995. Reprinted in G. Chaitin, *The Limits of Mathematics*, Springer-Verlag, 1998. Reprinted in G. Chaitin, *Thinking about Gödel & Turing*, World Scientific, 2007.

- G. Chaitin, "On the number of n-bit strings with maximum complexity," *Applied Mathematics and Computation* **59** (1993), pp. 97–100.

- G. Chaitin, "Responses to 'Theoretical mathematics...'," *Bulletin of the American Mathematical Society* **30** (1994), pp. 181–182.

- G. Chaitin, "Program-size complexity computes the halting problem," *Bulletin of the European Association for Theoretical Computer Science* **57** (Oct. 1995), p. 198.

- G. Chaitin, "The Berry paradox," *Complexity* **1**, No. 1 (1995), pp. 26–30.

- G. Chaitin, "A new version of algorithmic information theory," *Complexity* **1**, No. 4 (1995/1996), pp. 55–59.

- G. Chaitin, "How to run algorithmic information theory on a computer," *Complexity* **2**, No. 1 (Sept. 1996), pp. 15–21.

- G. Chaitin, "The limits of mathematics," *Journal of Universal Computer Science* **2** (1996), pp. 270–305. Reprinted in G. Chaitin, *The Limits of Mathematics*, Springer-Verlag, 1998.

- G. Chaitin, "An invitation to algorithmic information theory," in D. S. Bridges, C. Calude, J. Gibbons, S. Reeves, I. Witten, *Combinatorics, Complexity & Logic*, Springer-Verlag, 1997, pp. 1–23. Reprinted in G. Chaitin, *The Limits of Mathematics*, Springer-Verlag, 1998.

- G. Chaitin, *The Limits of Mathematics,* Springer-Verlag, 1998. Published in Japanese by SIB-Access, 2001.

- G. Chaitin, "Elegant LISP programs," in C. Calude, *People and Ideas in Theoretical Computer Science,* Springer-Verlag, 1999, pp. 32–52. Reprinted in G. Chaitin, *The Limits of Mathematics,* Springer-Verlag, 1998.

- G. Chaitin, *The Unknowable,* Springer-Verlag, 1999. Published in Japanese by SIB-Access, 2001.

- C. Calude, G. Chaitin, "Randomness everywhere," *Nature* **400** (1999), pp. 319–320.

- G. Chaitin, "A century of controversy over the foundations of mathematics," in C. Calude, G. Paun, *Finite vs. Infinite,* Springer-Verlag, 2000, pp. 75–100. Reprinted in Italian in V. Manca, *Logica matematica,* Bollati Boringhieri, 2001. Reprinted in G. Chaitin, *Conversations with a Mathematician,* Springer-Verlag, 2002. Reprinted in Turkish in B. S. Gür, *Matematik Felsefesi,* Orient, 2004. Reprinted in G. Chaitin, *Thinking about Gödel & Turing,* World Scientific, 2007.

- G. Chaitin, "A century of controversy over the foundations of mathematics," *Complexity* **5**, No. 5 (May/Jun. 2000), pp. 12–21. Reprinted in G. Chaitin, *Exploring Randomness,* Springer-Verlag, 2001. Reprinted in G. Chaitin, *Thinking about Gödel & Turing,* World Scientific, 2007.

- G. Chaitin, *Exploring Randomness,* Springer-Verlag, 2001.

- G. Chaitin, *Conversations with a Mathematician,* Springer-Verlag, 2002. Published in Japanese by Iwanami Shoten, 2003. Published in Portuguese by Gradiva, 2003.

- G. Chaitin, "Computers, paradoxes and the foundations of mathematics," *American Scientist* **90** (2002), pp. 164–171. Reprinted in G. Chaitin, *Meta Math!,* Pantheon, 2005.

- G. Chaitin, "Metamathematics and the foundations of mathematics," *Bulletin of the European Association for Theoretical Computer Science* **77** (Jun. 2002), pp. 167–179. Reprinted in Italian in *Enciclopedia del novecento: Supplemento III,* Vol. H-W, 2004, pp. 111–116. Reprinted in G. Chaitin, *Thinking about Gödel & Turing,* World Scientific, 2007.

- G. Chaitin, "Paradoxes of randomness," *Complexity* **7**, No. 5 (May/Jun. 2002), pp. 14–21. Reprinted in G. Chaitin, *Thinking about Gödel & Turing,* World Scientific, 2007.

- G. Chaitin, "Complexité, logique et hasard," in R. Benkirane, *La Complexité,* Le Pommier, Paris, 2002, pp. 283–310.

- G. Chaitin, "The unknowable," in A. Miyake, H. U. Obrist, *Bridge the Gap?,* Walther König, 2002, pp. 39–47.

- G. Chaitin, "Two philosophical applications of algorithmic information theory," in C. S. Calude, M. J. Dinneen, V. Vajnovszki, *Discrete Mathematics and Theoretical Computer Science,* Lecture Notes in Computer Science, Vol. 2731, Springer-Verlag,

2003, pp. 1–10. Reprinted in G. Chaitin, *Thinking about Gödel & Turing*, World Scientific, 2007.

- G. Chaitin, *From Philosophy to Program Size*, Tallinn Institute of Cybernetics, 2003.

- G. Chaitin, Interview with performance artist Marina Abramovic, in H. U. Obrist, *Interviews*, Charta, 2003, pp. 29–44.

- G. Chaitin, "L'univers est-il intelligible?", *La Recherche*, No. 370 (Dec. 2003), pp. 34–41. Reprinted in Italian in G. Chaitin, *Teoria algoritmica della complessità*, Giappichelli, 2006.

- G. Chaitin, "Thoughts on the Riemann hypothesis," *The Mathematical Intelligencer* **26**, No. 1 (Winter 2004), pp. 4–7.

- G. Chaitin, "On the intelligibility of the universe and the notions of simplicity, complexity and irreducibility," in W. Hogrebe, J. Bromand, *Grenzen und Grenzüberschreitungen*, Akademie Verlag, Berlin, 2004, pp. 517–534. Reprinted in G. Chaitin, *Meta Math!*, Pantheon, 2005. Reprinted in G. Chaitin, *Thinking about Gödel & Turing*, World Scientific, 2007.

- G. Chaitin, "Leibniz, information, math and physics," in W. Löffler, P. Weingartner, *Wissen und Glauben*, ÖBV & HPT, Vienna, 2004, pp. 277–286. Reprinted in Italian in G. Chaitin, *Teoria algoritmica della complessità*, Giappichelli, 2006. Reprinted in G. Chaitin, *Thinking about Gödel & Turing*, World Scientific, 2007.

- G. Chaitin, "Leibniz, randomness and the halting probability," *Mathematics Today* **40** (2004), pp. 138–139. Reprinted in Italian in G. Chaitin, *Teoria algoritmica della complessità*, Giappichelli, 2006. Reprinted in G. Chaitin, *Thinking about Gödel & Turing*, World Scientific, 2007.

- G. Chaitin, *Algorithmic Information Theory*, first paperback edition, Cambridge University Press, 2004. Identical to 4th printing.

- G. Chaitin, *Meta Math!*, Pantheon, 2005.

- G. Chaitin, "Algorithmic irreducibility in a cellular automata universe," *Journal of Universal Computer Science* **11** (2005), pp. 1901–1903.

- G. Chaitin, "The limits of reason," *Scientific American* **294**, No. 3 (Mar. 2006), pp. 74–81. Reprinted in G. Chaitin, *Thinking about Gödel & Turing*, World Scientific, 2007.

- G. Chaitin, "Probability and program-size for functions," *Fundamenta Informaticae* **71** (2006), pp. 367–370.

- G. Chaitin, "How real are real numbers?", *International Journal of Bifurcation and Chaos* **16** (2006), pp. 1841–1848. Reprinted in G. Chaitin, *Thinking about Gödel & Turing*, World Scientific, 2007.

- G. Chaitin, "Epistemology as information theory: From Leibniz to Ω," *Collapse* **1** (2006), pp. 27–51. Reprinted in Italian in G. Chaitin, *Teoria algoritmica della complessità*, Giappichelli, 2006. Reprinted in G. Chaitin, *Thinking about Gödel & Turing*, World Scientific, 2007.

- C. Calude, G. Chaitin, "A dialogue on mathematics and physics," *Bulletin of the European Association for Theoretical Computer Science* 90 (Oct. 2006), pp. 31–39.

- G. Chaitin, *Teoria algoritmica della complessità*, Giappichelli, Turin, 2006.

- G. Chaitin, *Meta Maths*, Atlantic Books, 2006. (UK edition of *Meta Math!*.)

- G. Chaitin, *Meta Math!*, first paperback edition, Vintage, 2006.

- G. Chaitin, "Is incompleteness a serious problem?", in G. Lolli, U. Pagallo, *La complessità di Gödel*, Giappichelli, 2007. Reprinted in G. Chaitin, *Thinking about Gödel & Turing*, World Scientific, 2007.

- G. Chaitin, "Speculations on biology, information and complexity," *Bulletin of the European Association for Theoretical Computer Science* **91** (Feb. 2007), pp. 231–237. Reprinted in G. Chaitin, *Thinking about Gödel & Turing*, World Scientific, 2007.

- G. Chaitin, "How much information can there be in a real number?", *International Journal of Bifurcation and Chaos* **17** (2007), in press. Reprinted in G. Chaitin, *Thinking about Gödel & Turing*, World Scientific, 2007.

- G. Chaitin, "The halting probability Ω: Irreducible complexity in pure mathematics," *Milan Journal of Mathematics* **75** (2007), in press. Reprinted in G. Chaitin, *Thinking about Gödel & Turing*, World Scientific, 2007.

- G. Chaitin, "Algorithmic information theory: Some recollections," in C. Calude, *Randomness and Complexity, from Leibniz to Chaitin*, World Scientific, 2007.

- G. Chaitin, "An algebraic characterization of the halting probability," *Fundamenta Informaticae* **79** (2007), in press.

- G. Chaitin, "The halting probability Ω: Concentrated creativity," in H. U. Obrist, *Formulas for the Twenty-First Century*, 2007. Reprinted in G. Chaitin, *Thinking about Gödel & Turing*, World Scientific, 2007.

Acknowledgements

- "On the difficulty of computations" is reprinted from *IEEE Transactions on Information Theory,* Vol. IT-16 (1970), pp. 5–9. Copyright © 1970 Institute of Electrical and Electronics Engineers, Inc.

- "Information-theoretic computational complexity" is reprinted from *IEEE Transactions on Information Theory,* Vol. IT-20 (1974), pp. 10–15. Copyright © 1974 Institute of Electrical and Electronics Engineers, Inc.

- "Randomness and mathematical proof" is reprinted from *Scientific American,* Vol. 232, No. 5 (May 1975), pp. 47–52. Reprinted with permission. Copyright © 1975 by Scientific American, Inc. All rights reserved.

- "Gödel's theorem and information" is reprinted from *International Journal of Theoretical Physics,* Vol. 21 (1982), pp. 941–954, with kind permission of Springer Science and Business Media.

- "Randomness in arithmetic" is reprinted from *Scientific American,* Vol. 259, No. 1 (July 1988), pp. 80–85. Reprinted with permission. Copyright © 1988 by Scientific American, Inc. All rights reserved.

- "Randomness in arithmetic and the decline & fall of reductionism in pure mathematics" is reprinted from *Bulletin of the European Association for Theoretical Computer Science,* No. 50 (June 1993), pp. 314–328, with the permission of the European Association for Theoretical Computer Science.

- "A century of controversy over the foundations of mathematics" (UMass-Lowell lecture) is reprinted from C. S. Calude, G. Paun, *Finite versus Infinite,* Springer-Verlag, 2000, pp. 75–100, with kind permission of Springer Science and Business Media.

- "A century of controversy over the foundations of mathematics" (Carnegie Mellon University lecture) is reprinted from *Complexity,* Vol. 5, No. 5 (May/June 2000), pp. 12–21. Copyright © 2000 Wiley Periodicals, Inc.

- "Metamathematics and the foundations of mathematics" is reprinted from *Bulletin of the European Association for Theoretical Computer Science,* No. 77 (June 2002), pp. 167–179, with the permission of the European Association for Theoretical Computer Science.

- "The halting probability Ω: Concentrated creativity" is reprinted from H. U. Obrist, *Formulas for the Twenty-First Century,* 2007. Copyright © 2007 G. J. Chaitin.

About the author

Gregory Chaitin is the discoverer of the remarkable Ω number, and has devoted his life to developing an information-theoretic approach to meta-mathematics and exploring its philosophical implications. He is a member of the Physical Sciences Department at the IBM Thomas J. Watson Research Center in Yorktown Heights, New York. He is also an honorary visiting professor in the Theoretical Computer Science Group at the University of Auckland (New Zealand), and an honorary professor at the University of Buenos Aires (Argentina).

Furthermore, Chaitin is a member of the *Académie Internationale de Philosophie des Sciences* (Belgium), a corresponding member of the *Academia Brasileira de Filosofia* (Rio de Janeiro), on the scientific advisory panel of the Foundational Questions in Physics & Cosmology Institute (FQXi), the honorary president of the scientific committee of the *Instituto de Sistemas Complejos de Valparaíso* (Chile), and a permanent member of the Rutgers University Center for Discrete Mathematics & Theoretical Computer Science (DIMACS). He has an honorary doctorate from the University of Maine.

This is Chaitin's eleventh book.